The Language of Life

Also by Francis S. Collins

The Language of God: A Scientist Presents Evidence for Belief

The Language of Life

DNA AND THE REVOLUTION IN PERSONALIZED MEDICINE

Francis S. Collins

HARPER

An Imprint of HarperCollins*Publishers*
www.harpercollins.com

FIRST EDITION

Designed by Joy O'Meara

Library of Congress Cataloging-in-Publication Data

Collins, Francis S.

The language of life : DNA and the revolution in personalized medicine / Francis S. Collins.—1st ed.

   p. cm.

ISBN 978-0-06-173317-8

1. Genetic screening—Popular works. 2. Medical genetics—Popular works. I. Title.
[DNLM: 1. Genetic Diseases, Inborn. 2. Genetic Counseling. 3. Genetic Screening. 4. Genetics, Medical. QZ 50 C712L 2009]

RB155.65.C65   2009
616'.042—dc22   2009025832

---

10 11 12 13 14 ov/rrd 10 9 8 7 6 5 4 3 2 1

To my many wonderful teachers: mentors, colleagues, patients, and families

Acknowledgments

I am deeply grateful to the many selfless individuals and families who allowed me to include their personal stories in this book, providing the reader with important insights into the realities of personalized medicine. That list includes Blake Aldhaus and his mother, Anita; Tracy Beck; Sam Berns and his parents, Scott Berns and Leslie Gordon; Sergey Brin and Anne Wojcicki; McKenzie Christeson and her parents, Scott and Kris Wood; Bill Elder; Marvin Frazier; Doris Goldman; Jeffrey Gulcher; Wayne Joseph; Judy Orem; Kate Robbins; Anabel Stenzel and Isabel Stenzel Byrnes; Dale Turner; and several others who chose not to have their names used.

I am also greatly indebted to a long list of busy experts who took the time to read a draft of parts or all of the manuscript and provided valuable feedback on its content. That includes Melissa Ashlock, Barbara Biesecker, Stephen Chanock, Marc Chevalier, Brandon Collins, Tony Fauci, Greg Feero, David Ginsburg, Alan Guttmacher, Richard Hodes, Kathy Hudson, Tom Insel, Mark Kay, Teri Manolio, Judy Mosedale, Sharon Terry, Larry Thompson, Eric Topol, and Dick Weinshilboum. Their collective suggestions made this book substantially more accurate and more accessible than it otherwise would have been, but I take full responsibility for any errors that remain.

Thanks go to Judy Hutchinson, who carefully transcribed the dictated first drafts of each chapter; to my wife, Diane Baker, who provided valuable input on content and keyed in all of the initial edits;

and to biomedical illustrator Darryl Leja, who ably and elegantly rendered all of the artwork. Finally, I want to thank my publishing agent, Gail Ross, for her unflagging support; and my editor, Bruce Nichols, for believing in the importance of this book even before I did, for encouraging me to make it a reality, and for providing wise perspective all along the way.

Contents

We're Not in Kansas Anymore

The tearful young man was on the phone with his uncle. "My mother is dying. She's in a coma, and I don't think she'll make it through the night." Surrounded by the drone of whirring centrifuges and students' conversation about the previous night's lab party, Dr. Robert James moved to a quiet spot where he could speak privately to his distraught nephew.

"I'm so sorry, Brad," he said. "Your mother truly fought a valiant battle against ovarian cancer, and she rallied so many times when it seemed that all was lost. But it sounds as if this is really the end. What can I do to help?"

"Well," said his nephew, "my sister and I have been worrying about whether her cancer might be hereditary, given all of the other women on her mother's side who suffered from breast cancer or ovarian cancer. You once told us that someday there might be a test to determine whether one or both of us had inherited her cancer risk. If that's true, is it too late to pursue this?"

Dr. James explained how to proceed. A blood sample from his dying sister-in-law was shipped to Dr. James's lab the next day, DNA was prepared, and the sample was carefully stored in the freezer. He thought it was unlikely this would ever be useful, but at least it was something to do.

Five years later, Brad's sister Katherine contacted Dr. James, explaining that she had been reading articles in the popular press about the discovery of genes involved in hereditary breast and ovarian cancer. Katherine had been having yearly mammograms, even though she was only in her thirties, but she was particularly concerned that there were no good screening tests available for detecting early ovarian cancer. Her mother had originally been diagnosed with this cancer at 52, and Katherine thought every day about her own potential for developing it.

Dr. James confirmed that the discovery of genes known as *BRCA1* and *BRCA2* might well make it possible to be more precise about the risk of cancer in the family, if it turned out that Katherine's mother carried a mutation in one of these genes. Concerned about the risk of losing her health insurance if she tested positive, Katherine wanted to know whether there was some other way to get the information. Her uncle told her about a clinical research study in a nearby city that allowed testing under an assumed name, and Katherine decided to proceed. After genetic counseling about the risks of knowing or not knowing this information, Katherine requested that the DNA sample on her mother, carefully stored for several years in the freezer in Dr. James's laboratory, be forwarded to a testing facility.

A few weeks later Katherine called Dr. James to report that a significant *BRCA1* mutation had been found in her mother's DNA. Katherine faced a 50 percent risk of having inherited that misspelling, in which case her lifetime risk of breast cancer would be approximately 80 percent, and that of ovarian cancer about 50 percent. Katherine was deeply concerned about herself, but even more so about her six-year-old daughter.

She spent two weeks waiting for her own results, and it seemed like an eternity. She tried to imagine what she would do with a positive result. Would she approach a surgeon about removing her ovaries? Would she even contemplate removing both breasts and undergoing surgical reconstruction, as many women with mutations in *BRCA1* or *BRCA2* have done? What would she tell her daughter, and at what age should her daughter be tested? Some days she was certain the test would be positive—after all, everyone remarked how much she looked like her mother. On other days, she remembered that such information was irrelevant to the possibility of her carrying this specific genetic glitch, and she was more hopeful.

The fateful day arrived when a phone call from the genetic counselor invited Katherine to come to the clinic to hear the results. With her heart in her mouth, she sat across the desk as the counselor opened the file and then broke into a smile. "Katherine," she said, "I have good news. You have not inherited the *BRCA1* mutation carried by your mother. Your risk of breast and ovarian cancer is no greater than that of the average woman of your age, and your daughter likewise carries no special risks for these diseases."

Overjoyed, Katherine called her uncle to share this happy moment. But both of them confessed to remaining uneasy about other maternal relatives in Canada and Europe, and about Katherine's brother, Brad, who had chosen not to be tested. Although males with mutations in *BRCA1* and *BRCA2* face only a slightly increased risk of cancer of the prostate, pancreas, and male breast, their daughters may still be at high risk of breast and ovarian cancer if they've inherited the mutation. Brad's young daughter now became the remaining member of this nuclear family with a potential genetic cloud over her.

Dr. James is a physician who has devoted his professional life to research on molecular genetics, so it was ironic that his own family turned out to be affected by one of the more dramatic discoveries in hereditary disease of the past decade.

But then it happened again. This time, it was his father-in-law, Fred, now in his late seventies, who contacted Dr. James about a medical evaluation. Fred had noticed some discomfort in his legs and a deterioration in his golf game and, after an initial evaluation by his primary physician, had been referred to a neurologist.

Fred was calling to say that the neurologist had detected some slowing of nerve conduction in his legs, and was suggesting that Fred should be tested for an uncommon genetic condition known as Charcot-Marie-Tooth disease, named for the three French investigators who originally identified it. Dr. James was initially appalled at the idea of such testing, since Charcot-Marie-Tooth disease was generally associated with progressive weakness in the legs beginning in the twenties and thirties. Thinking that a genetic test for this condition in an elderly man would be essentially a waste of time and money, Dr. James nonetheless did not voice an objection to this plan, since he didn't want to interfere with his father-in-law's medical evaluation. To his amazement and consternation, the test was positive. After more study of the problem and discussion with the experts, it began to make more sense. Until DNA testing was made available, Charcot-Marie-Tooth disease had been purely a clinical diagnosis. So of course the cases that were discussed in textbooks and medical journals tended to be those with a more severe course. Now that the gene had been identified and could be spotted by a specific molecular test, it was becoming apparent that a milder disease, including the remarkably late onset presented by Fred, was more common than had been appreciated.

This time the diagnosis struck even closer to home. Charcot-Marie-Tooth disease is a dominant condition, and this means that the child of an affected individual has a 50 percent chance of inheriting the abnormal gene and also being affected. Thus Dr. James's wife, Dawn, as well as her brother and sister, might be significantly affected by this discovery in the future. In fact, it wasn't just a matter of the future; it was also about the past and the present. Dawn's sister, Laura, had long struggled with what had been assumed to be a congenital problem with her feet and ankles. Blamed on "club feet," but never definitively diagnosed, this problem now appeared likely to be a consequence of a particularly early manifestation of the same genetic disease that had appeared so late in her father. Here was a chance to provide a definitive diagnosis. Yet Laura decided not to be tested. She was not convinced that the information would change anything, and she was a cynic about health care. She had had several frustrating experiences over the years with orthopedic interventions that were supposed to help her chronic foot problems but didn't really provide much relief. She respected her brother-in-law, Dr. James, but not the system.

For her part, Dawn considered the possibility of testing, even though she had no symptoms of this disease and was now in her mid-fifties. She ultimately decided to embrace the ambiguity of the situation, rather than obtain a definitive answer, as she was not sure how a positive test result would change her outlook. Dr. James was somewhat puzzled, but he supported her decision. After all, she was healthy and happy. By contrast, he wished he could change her sister's mind. Shouldn't Laura know why she had suffered so long?

Robert James happens to be an M.D. and a geneticist. How surprising is it that he faced two situations involving genetic testing and risk in his own family? Actually, not very. The National Organization for Rare Diseases (NORD) estimates that there are at least 6,000 rare (so-called orphan) diseases, defined as conditions that affect fewer than 200,000 people in the United States. Collectively, 25 million Americans are affected by one of these conditions. If you include their families and friends, then few of us have not been touched in some way by one of these conditions. Many of them are caused by genes that have acquired misspellings somewhere in the family, as in the cases confronted by Dr. James.

So rare diseases are not so rare after all. In fact, I can now reveal the truth. "Dr. James" is a pseudonym: for myself. I've changed all the other names, but Brad and Katherine are my nephew and niece, Fred is my father-in-law, Laura is my sister-in-law, and Dawn is my wife. Many years ago I entered the field of genetic medicine with the hope of contributing something to the understanding of other people's medical issues and other families 'challenges. In fact, genetic medicine has brought the problems of rare genetic conditions right to my own door.

Discoveries about genetics are not limited to just those 6,000 conditions of a strongly hereditary nature, however. We are now in the midst of a genetic revolution that will touch all of us in numerous ways: this revolution involves common diseases like diabetes, heart disease, cancer, asthma, arthritis, Alzheimer's disease, and more; mental health and personality; decisions about child bearing; and even our ethnic histories. We now see that the language spoken by our DNA is the language of life itself. And we are now reading this language in ways that could have a profound effect on your health.

An explosion of research in the last few years has taken us from a general observation that diseases tend to run in families to the discovery of very precise DNA variations that play a predictable role in many diseases, and that can be used to make increasingly accurate predictions about an individual's potential future likelihood of illness. No longer does this apply just to rare diseases like Charcot-Marie-Tooth or the specific breast cancer caused by *BRCA1* mutations. A veritable deluge of discoveries of DNA glitches that play a role in risks of common disease has been pouring out of leading laboratories around the world and shows no likelihood of tapering off in the near future. We have crossed the threshold from a medical model where genetic testing was possible only in a high-risk situation, as in the two examples in my family, to circumstances in which such tests are being offered to virtually anyone.

These developments have electrified the public and are now regularly reported in the media and featured on *Oprah*. Companies market complex DNA analysis directly to the public, arguing that the time has come for prevention-minded individuals to take the tests and be empowered by this specific information. One of these companies, 23andMe, named for the 23 pairs of human chromosomes, urges potential customers to "unlock the secrets of your DNA today." A competing company, Navigenics, claims that its testing "gives you action steps to take control of your health." A third company, deCODE, says that its testing service "enables you to make more informed decisions about your health."

Currently, these tests sample less than one-tenth of 1 percent of the complete DNA molecule, but the information they yield bears on dozens of diseases and conditions. These numbers will expand rapidly in the near future. Discoveries are announced almost weekly, as we unlock the secrets of the rest of the genome.

I have already revealed a lot about my own family's experience with specific genetic conditions. But that knowledge came from doctor-driven, specialized testing. What about the knowledge available to each of us today from these new providers of direct access to DNA information? As your guide to this new era of personalized medicine, I asked myself whether it would really be appropriate to stand on the sidelines when it comes to this groundbreaking new era in comprehensive DNA analysis. Or should I do the genetic equivalent of the "full Monty"? As recently as two years ago, I would have concluded that the ability to make meaningful comprehensive predictions about future illness from DNA testing was premature. But now, the landscape is undergoing rapid change. Knowing full well that these are early days for making accurate predictions, I still decided it was time to find out. I conferred with my adult daughters, since this kind of testing might also reveal things about them, and they encouraged me to "go for it."

Family medical history is of course a critically important guide. I am blessed with a remarkably healthy group of close relatives—both my parents lived to age ninety-eight, and my three brothers (all older than me) are all athletic and in excellent health. So my own likelihood of future illness is hard to discern from my pedigree. But could there be risks lurking in my DNA that have not shown themselves?

Besides this curiosity about my own genome, I was also interested to find out how these direct-to-consumer companies conduct business and report results. Is their laboratory work accurate? How do they convert a DNA result into a prediction about risk? And how good are they at conveying that information in a fashion that empowers rather than confuses the consumer?

I decided to submit a DNA sample to each of the three companies offering comprehensive DNA analysis. (There are quite a number of other companies—some credible, some not—that are more focused on specific tests for specific purposes.) I decided not to use my own name, as I didn't want these companies to treat me any differently than they would a typical customer.

The costs of the test were substantially different: 23andMe charged just $399, whereas deCODE cost $985, and Navigenics charged $2,499 (but offered telephone genetic counseling as an added feature). The DNA sampling processes were easy: spitting into a special tube for 23andMe and Navigenics, and scraping my cheek for deCODE. Each company promised confidentiality through assignment of passwords to their Web site. And while there were some interesting differences, the lists of conditions tested were heavily overlapping (see Appendix E).

23andMe was the first to report results, in just two weeks. de-CODE weighed in a couple of weeks later, and Navigenics reported after seven weeks (but strangely they had not actually completed the analysis, and 7 of the 25 conditions being tested for still had some results pending). As much as I knew the significant limitations of the tests to make precise predictions, I still found it both exciting and a bit unnerving to enter my password and begin to review my own results. Each Web site was reasonably well designed to help me understand the results and to put my own risk in context of the average person. Of the three, I found the 23andMe Web site to be most user-friendly.

To assess genetic risk, all three companies base their work on the same publications in the scientific literature. So in many instances they tested exactly the same variants in my DNA. I looked closely at the details to see if any of the actual lab results were discordant. To my relief, I couldn't find a single example where that was so. So the actual DNA analysis is apparently of very high quality.

What did I learn? For most common diseases, I was happy to see that my risk scored as average or below average. But there were some significant exceptions. All three companies agreed that my risk for type 2 (adult-onset) diabetes was elevated. Though the precise risk estimate varied slightly, my risk came in at about 29 percent, somewhat higher than the average person (23 percent). My risk of age-related macular degeneration, a common cause of blindness in the elderly, and which had taken my aunt's eyesight in her eighties, was also substantially higher than that of the average person. And the chance that I would be affected by a particular type of glaucoma was also elevated, though the companies disagreed about the absolute risk.

Of course this was all statistical information—there was no proof that I would definitely get any of these diseases, and the predictions didn't take into account my family medical history at all. But despite my being aware of all the shortcomings of these tests, the information had an immediate effect on my view of the future. As a physician, I had known for years about a long list of general recommendations for maintaining good health, but I hadn't necessarily followed them. Now, with these specific threats, I found I was more attentive. Even though the predicted 29 percent risk of diabetes was marginally higher than the 23 percent baseline, and even though my negative family history and absence of obesity no doubt reduced my risk even further, I resolved to go ahead with a long-postponed plan to contact a personal trainer and work harder at a diet and exercise program, knowing that this was the best prevention for whatever diabetes risk still remained. I looked up the most recent research articles on macular degeneration, and concluded that the evidence supporting the protective effect of omega-3 fatty acids was solid enough that it would be a good idea for me to include more fish in my diet. And given the glaucoma risk, I resolved to be sure to have my eyes checked each year, including measurement of intraocular pressure. Were these all things I should have been doing anyway? Perhaps. But we are constantly bombarded by all kinds of generic health advice—eat fish! take a daily aspirin! drink red wine! exercise!—and it's hard if not impossible to remember to do all these things. Despite all of the limitations of the data, the disclosure of this personalized genetic information provided a motivator for specific actions.

There was one test result I thought seriously about just not looking at—the one for Alzheimer's disease risk. This is one of the strongest genetic risk factors yet identified, capable of increasing one's risk by as much as eightfold. And at the present state of medical research, there is nothing you can do about it, other than use the information to try to plan for the future. There's no convincing evidence that diet or medication will delay or prevent the onset of Alzheimer's disease in a susceptible person. Despite my negative family history for Alzheimer's disease, I felt my heart rate go up as I decided to click on the button and reveal the result. The answer was a relief—my lifetime risk of Alzheimer's disease comes out lower than average, at just 3.5 percent.

A few other results caught my eye. 23andMe and deCODE reported on my ability to metabolize a commonly used drug for blood clots, called coumadin. I have never taken that drug, but my mother was on it for several years and proved to be unusually sensitive, so that her dose had to be adjusted downward to avoid toxicity. Sure enough, the 23andMe report predicted that I would also have "increased sensitivity." Oddly, deCODE looked at exactly the same variants in my DNA, got the same results, but predicted I would need an "average dose." This was a good reminder of the immature state of making predictions from these DNA results. These companies are all looking at the same scientific evidence, but regrettably they haven't achieved consensus on interpretation. They should get together urgently to do this, or the public may start to become confused and potentially disillusioned.

This discordance between the results from the three companies was most apparent for the prostate cancer risk prediction. My father had this disease late in life, and so when my 23andMe results arrived, I was relieved to see a prediction of lower than average risk. But then deCODE disagreed, saying my risk was slightly elevated. Navigenics upped the ante substantially, placing me at a 40 percent higher risk than the average male (24 percent compared to a baseline of 17 percent). What on earth was going on here? To sort this out, I had to drill down into the details of the lab studies—and I discovered the explanation. 23andMe had tested for just 5 variants known to confer prostate cancer risk; deCODE had tested for 13; and Navigenics had tested for 9. There was considerable overlap between the DNA markers tested, but no company had actually tested for the complete set of 16. Having all the results in front of me, I could calculate that risk, and it came rather close to the Navigenics prediction. So the reassurance I had first obtained from 23andMe was short-lived—here was another condition that I should pay close attention to.

There's a really important lesson here—the field is moving so quickly that any genetic risk predictions based on today's understanding will need to be revised in the context of new discoveries tomorrow. That applies not just to prostate cancer but to all of the rest of my risk predictions—what is possible now is only a blurry picture of reality. As genetic tests get better, and other critical information such as family medical history and current medical status get more effectively integrated with the DNA results, the picture will come increasingly into focus. So anyone embarking on this adventure should be prepared to revisit the risk estimates on a regular basis as new knowledge is obtained.

As the most expensive of the three options, and the one most focused on medical applications, Navigenics also offered the chance to consult with a genetic counselor about my results. I spoke to one of their counselors on the phone, playing my role as an interested consumer without much scientific training. The counselor was careful to say she was not dispensing medical advice, but after going over my DNA results she strongly recommended that I see a physician about my prostate cancer risk. I expressed concern that my physician might not know what to make of these genetic tests, and she indicated that lots of physicians were now calling Navigenics for advice. I asked whether these DNA-based predictions might change in the future, and she correctly pointed out that new information was being derived every day, and Navigenics would keep me informed by e-mail as the predictions became more refined. Oddly, however, she implied that most of the remaining genetic risk factors for common disease will have been discovered in the next two or three years; as a scientist working in this field, that seems quite unlikely to me.

23andMe also included a section to detect carriers of conditions that should not affect my own health, but might put my children at risk if their mother was also a carrier and they were unlucky enough to inherit the misspelled gene from both of us. I learned that I am a carrier of two recessive diseases that cause adult-onset medical problems—alpha-1-antitrypsin deficiency and hemochromatosis. The former can lead to emphysema and/or liver disease, the latter to a buildup of body iron that can result in cirrhosis, heart failure, and diabetes, among other serious conditions. Those results led to a conversation with my daughters, who were clearly concerned to hear that specific defective genes were traveling through the family. Though we all knew that this must be the case, having the specific culprits identified made this a much more concrete situation and has led both of my daughters to begin to explore testing for themselves.

23andMe also provided results for several nonmedical traits. Predictions of wet ear wax, ability to taste bitter foods like Brussels sprouts, and so on were entertaining—but the limitations of this kind of testing were at once apparent when the Web report predicted that I have brown eyes (they are quite definitely blue).

Both 23andMe and deCODE also provided information about my likely ancestry. I had secretly hoped for some exotic African, Asian, or Native American revelations, but there were no real surprises—I seem to be rather monotonously European, except for a tiny blip on the eighth chromosome that looks Asian in origin.

So that's my full Monty, at least as far as the technology currently allows. But whether or not you choose to get your own test done now, I am here to tell you that you may not be able to avoid it much longer.

We are on the leading edge of a true revolution in medicine, one that promises to transform the traditional "one size fits all" approach into a much more powerful strategy that considers each individual as unique and as having special characteristics that should guide an approach to staying healthy. Although the scientific details to back up these broad claims are still evolving, the outline of a dramatic paradigm shift is coming into focus.

The analysis I had done tested one million places in my DNA. But this is just the beginning. Soon, probably within the next five to seven years, each of us will have the opportunity to have our complete DNA sequenced, all three billion letters of the code, at a cost of less than $1,000. This information will be very complex but also very powerful. Careful analysis of the complete content of your genome will allow a considerably more useful estimate of your future risks of illness than is currently possible, enabling a personalized plan of preventive medicine to be established.

Many people, when first confronted with the chance for such foreknowledge, say, "I don't want to know; it's better to enjoy life than worry about future risks." They might agree with the blind seer Tiresias in Sophocles' play about Oedipus, who was doomed to see the future but to be unable to change it; Tiresias lamented that "it is but sorrow to be wise when wisdom profits not." But we are not as powerless as Tiresias. In many cases of predicting genetic risk, this kind of wisdom may well provide personal health benefits. As I experienced myself, learning your own DNA secrets can be the best strategy for protecting your health and your life.

This opportunity to empower each of us to practice better prevention and treatment is not just about DNA. Studies of the interaction between genetic and environmental risks are pinpointing critical parts of our health that derive from environmental variables. That will lead to a greater opportunity for you to monitor and adjust your environmental exposures to improve your chances of staying well or recovering from an illness.

Your DNA sequence, properly encrypted, will soon become a permanent part of your electronic medical record, and will be utilized by health care professionals to make a wide variety of decisions about drug prescriptions, diagnostics, and disease prevention. If you fall ill, the therapeutic options waiting for you, many derived from new understanding of the human genome, will be both more effective and less toxic than the treatments available just a few years ago. Many of these therapies will be in pill form, but some will be gene therapies, in which the gene itself becomes the drug. Some will even be cell therapies, based upon the ability to take your own skin cells or blood cells and transform them into cells you might need in, say, your pancreas for diabetes, or your brain for Parkinson's disease.

This book is a report from the front lines of a revolution. It is also a user's manual of what you need to know to benefit your own health, and that of your family. There are things you can do right now— starting with a family medical history—to prepare. But first, you have to be ready to embrace this new world.

For centuries, we considered ourselves to be healthy until symptoms of illness arose. Once diagnosed, correctly or not, we received standardized treatments. In accordance with this view, the human body was generally ignored until something went wrong.

Today, we have discovered that everyone is born with dozens of genetic glitches. There are no perfect human specimens. But not all our glitches are the same, so one treatment often does not fit all sufferers of a given disease. Not just our medicine but our fundamental attitude toward the human body is changing.

A lot has been written—often breathlessly—about the DNA revolution. But this book aims to be about the facts. This is a book about hope, not hype. The accelerating ability to read the language of life is allowing a completely new view of health and disease. If you are interested in living life to the fullest, it is time to harness your double helix for health and learn what this paradigm shift is all about.

The Language of Life

CHAPTER ONE

The Future Has
Already Happened

Scientists aren't generally prone to effusiveness. We are privately excited about our work, but in public we often, and rightly, emphasize skepticism and caution. But there are exceptional moments where skepticism is set aside, electricity fills the room, and a scientist with palpable passion and flashing eyes describes unabashedly a change in the landscape that will have lasting significance.

Just five months into the new millennium, I had that experience. Together with more than 2,000 of my colleagues, laboring in 20 centers in 6 countries, we had succeeded in reading almost 90 percent of the letters of the human *DNA* instruction book, otherwise known as the human *genome*. After much anticipation, and many tumultuous moments, the achievement of an almost impossibly audacious goal that had motivated all of us for a decade was now essentially assured.

The public announcement of the complete draft of the genome would follow one month later at the White House. But on this Saturday in May 2000, it fell to me, as the "field marshal" of the international Human Genome Project, to deliver the keynote address at the annual gathering of the genome science community, held at Cold Spring Harbor Laboratory on Long Island. This was the private, science-only version of the coming White House announcement. Presided over by James Watson, who with Francis Crick had discovered the double helical structure of DNA in 1953, Cold Spring Harbor was the Mecca where genome scientists made their pilgrimages every year.

But 2000 was a year like no other. Looking around me at the faces of so many scientists, both young and old, who had worked together to achieve this historic goal, I began my presentation: "We have been engaged in a historic adventure. Whether your metaphor is Neil Armstrong or Lewis and Clark, your metaphor is at risk of falling short. There is no question that the enterprise we have gathered here to discuss will change our concepts of human biology, our approach to health and disease, and our view of ourselves. This is the moment, the time when the majority of the human genome sequence, some 85 percent of it, looms into view. You will remember this. You will tell your future graduate students, perhaps even your future grandchildren, that you sat, stood, or sprawled in Grace Auditorium, in the presence of the intellectual giants of genomics that fill this hall right now, and of Jim Watson himself, and reflected upon this astounding time in our history." (A personal history of the Human Genome Project can be found in Appendix C.)

Everyone in the room knew that the science of DNA had reached an inflection point. With the sequencing of the entire genome, scientists could launch into a dizzying array of groundbreaking research projects to unlock the greatest secret of the human body. How does our DNA, life's instruction manual, actually work? We had climbed to the top of one big mountain, and we were about to start rushing down the other side, into a valley full of potential discoveries.

THE GENOME REVOLUTION

Nearly a decade has passed since that moment of celebration. Virtually all biomedical researchers would agree that their approach to understanding how life works has been profoundly and irreversibly affected by access to the complete DNA sequence of the human genome, and that of many other organisms. Graduate students probably cannot imagine how anybody did research in human genetics without having this information available at the click of a computer mouse.

But the effect on the public of all the hoopla in 2000 has been mixed. Most people know that the genome has been spelled out, but they have lost track of what has happened since then. They remember the ascent of the mountain, but they are unaware of the rewards that are starting to appear in the valley. Some of the press announcements at the time implied an immediate transformation of medicine, but that was never realistic—lead times between basic discoveries in science and changes in practical medicine, technology, or daily life tend to be measured in decades. Indeed, most of the promise offered by the sequencing of the human genome still lies ahead. But the leading edge has arrived, and is already affecting many lives. For starters, let's consider the very real case of Karen, diagnosed with breast cancer in 2005.

Karen Vance (not her real name) was just 40 years old when she found a lump in her breast. A mammogram was negative, but ultrasound revealed a 2-centimeter mass, and a biopsy revealed breast cancer. After consultation with her surgeon, she underwent a lumpectomy and removal of 23 lymph nodes, all of which were negative for cancer. Since her mother had also had breast cancer (though not until age 64), and there was also a history of breast cancer on her father's side, Karen decided to pursue testing for the *BRCA1* and *BRCA2* genes, but those tests were negative. She went through the usual radiation treatments following the lumpectomy, and then was faced with deciding whether to proceed with chemotherapy to reduce the risk of a future recurrence. She consulted with no fewer than three oncologists, all of whom recommended aggressive chemotherapy because of her relatively young age. Karen struggled with that recommendation, but given the unanimity of medical opinion, she resigned herself to proceeding with chemotherapy. She began to explore what kind of wig she might need to purchase.

To her great surprise, she received a phone call from her brother, who was not a medical professional but had seen a story on television about a new test that could be done on the actual breast cancer tissue to make a more precise prediction about the likelihood of recurrence. This test was based on an analysis of which genes were turned on or off in the tumor. Validation on thousands of cases had shown that this "gene expression" analysis gave a more precise picture of that tumor's likely aggressiveness than the traditional approach of looking through a microscope at the appearance of the cells.

Karen consulted with her surgeon, who had heard of the test, though he had not had much experience with it. He agreed to send off her tumor sample to the testing laboratory. Just four days before she was due to start chemotherapy, the test came back, indicating a very low recurrence score. One of her three oncologists was skeptical, but the other two were convinced that this information provided justification for pursuing hormonal therapy alone. Karen decided to adopt that approach. Four years later, she remains completely free of any signs of disease. This particular test represents one of the first fruits of the genomic revolution, and Karen is one of its pioneers.

Karen's case exemplifies a new approach to medicine that will soon affect virtually all facets of health care. No longer satisfied with empirical or superficial explanations of disease, scientists are peering into the molecular basis of cancer, heart disease, diabetes, Alzheimer's disease, schizophrenia, autism, and virtually all other conditions, peeling off the layers of the onion and finding that many accepted principles of medicine and biology require substantial revision. Fundamental gaps in our understanding about the human body are now being filled in. Hereditary factors for nearly all diseases are now being pinpointed as specific glitches in DNA, and these are appearing in great numbers following the completion of the Human Genome Project. As a result, healthy individuals are increasingly able to discover some of their body's inner secrets and take appropriate action. The potential for individual prediction is beginning to spill out to the general public, offering the opportunity to take more control of your fate.

Those who develop a disease, like Karen, are now offered molecular tools to predict the course, or even to decide that therapy isn't necessary. And the range of therapeutic options is expanding, as knowledge about the human genome provides new targets for the development of powerful treatments. None of this is happening overnight, and ultimate success will depend upon the visionary investment of energy, talent, and financial resources by scientists, governments, universities, philanthropic foundations, biotechnology and pharmaceutical companies, and the general public. But without question, man's knowledge of man is undergoing the greatest revolution since Leonardo.

DNA IS THE LANGUAGE OF LIFE

The discoveries of the past decade, little known to most of the public, have completely overturned much of what used to be taught in high school biology. If you thought the DNA molecule comprised thousands of genes but far more "junk DNA," think again. If you thought the human genome must be the most complex version of DNA on earth, think again.

For the purposes of this book, there is no need to learn every detail of DNA structure (for some of that, see Appendix B). This book is about applications, not engineering. But to understand those applications, it is important to learn some of the principles and some of the vocabulary.

Bacteria have DNA. Yeast have DNA. So do porcupines, peaches, and people. It is the universal language of all living things. We are in a truly historic era, when this language from many different species is being revealed for the first time. All of the DNA of an organism is called its *genome*, and the size of the genome is commonly expressed as the number of *base pairs* it contains. Think of the twisted helix of DNA as a ladder. The rungs of the ladder consist of pairs of four chemicals, called bases, abbreviated *A, C, T, G.* As shown in Figure 1.1, DNA is a long ladder. Its backbone is a monotonous string of sugars and phosphates. The information content resides in those chemical bases arranged within the interior, where A always pairs with T, and C always pairs with G. The simplest free-living single-cell organisms, such as bacteria, generally pack all their information into a genome of a few million base pairs. Fancier multicellular organisms with more complex body plans require larger genomes to specify those functions. Our own genome stacks up as 3.1 billion rungs of the DNA ladder. Most other mammals have genomes of about that size, give or take a billion or so, but many amphibians have genomes substantially larger than ours, and a very simple plant called the whisk fern, lacking flowers, fruit, or even leaves, has a genome 100 times larger than our own!

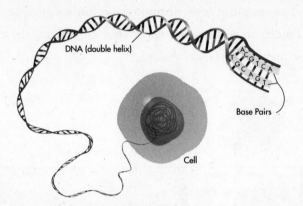

Figure 1.1: The double-helical DNA information molecule, the "instruction book" of all living things, here shown spilling out of the nucleus of a cell. The information content of DNA is specified by the order of the chemical bases (A, C, G, or T). Each of the two strands carries the complete information, since A always pairs with T, and C always pairs with G.

A *gene* is a segment of the DNA ladder that carries a packet of functional information. The shortest genes are only a few hundred base pairs in length; the longest, the Duchenne muscular dystrophy gene, stretches to more than 2 million rungs of the ladder. The best-understood genes are those that code for *protein*. This process involves first making an *RNA* copy of the DNA; that RNA is then transported to the ribosome "protein factories" in the cytoplasm, where the letters of the RNA code are translated into the amino acids used by proteins (Figure 1.2). This translation is carried out using a triplet code word; for example, AAA in the RNA codes for the amino acid lysine, and AGA codes for arginine. Mistakes in the DNA will lead to mistakes in the RNA, and that can result in garbling of the protein (Figure 1.3).

One major surprise from the Human Genome Project was the discovery that human DNA contained only 20,000 protein-coding genes. We expected a lot more than that! Even the lowly roundworm has about 19,000 genes. Some observers were actually upset by this apparent downgrading of our importance, but we assume our genome must be fancier in some other way. After all, we are the only species that has sequenced our own genome!

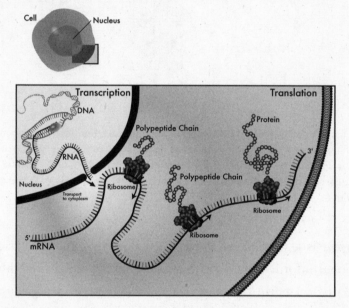

Figure 1.2: The "basic dogma of molecular biology": DNA codes for messenger RNA (called transcription, takes place in the nucleus), which then codes for protein (called translation, carried out by ribosomes in the cytoplasm).

Normal	DNA	HAS	ALL	YOU	CAN	ASK	FOR
Missense	DNA	HAS	ALL	LOU	CAN	ASK	FOR
Nonsense	DNA	HAS	ALL	YOU	STOP		
Frameshift	DNA	HAS	ALY	OUC	ANA	SKF	OR

Figure 1.3: Translation from RNA to protein occurs using a word length of three letters. Mistakes in the DNA genome lead to mistakes in messenger RNA, which then lead to mistakes of various types in the protein.

Most genes are internally interrupted in a puzzling way by long stretches of DNA information that are removed, or "spliced out," in the process of creating a mature RNA that is ready to be translated into protein. On the average, each gene has eight of these removable sequences, called *introns*. These interrupt the actual coding segments, called *exons*. Depending on which of the introns and exons are removed, and in what order, a given gene can produce several different proteins. Imagine a gene that has the basic structure CiAiUiRiEiT, where each of the capital letters represents an exon and i represents the introns. Depending on the splicing pattern, this gene could make CARE, CURE, CAR, CART, CAT, CUT, ARE, ART, and even CURiE (if one of the introns was retained).

The exons and introns of protein-coding genes add up together to about 30 percent of the genome. Of that 30 percent, 1.5 percent are coding exons and 28.5 percent are removable introns. What about the rest? It appears there are also long "spacer" segments of DNA that lie between genes and that don't code for protein. In some instances, these regions extend across hundreds of thousands or even millions of base pairs, in which case they are referred to rather dismissively as "gene deserts." These regions are not just filler, however. They contain many of the signals that are needed to instruct a nearby gene about whether it should be on or off at a given developmental time in a given tissue. Furthermore, we are learning that there may be thousands of genes hanging out in these so-called deserts that don't code for protein at all. They are copied into RNA, but those RNA molecules are never translated—instead, they serve some other important function.

Our genome is littered with repetitive sequences that have been inserted during a series of ancient assaults by various families of DNA parasites. Once they gain access to the genome, these "jumping genes" are capable of making copies of themselves, and then inserting those copies randomly throughout the genome. Roughly 50 percent of the human genome has had this history. However, in a nice demonstration of how natural selection can operate on all sorts of opportunities, a small fraction of these jumping genes have actually landed in a place where they have provided some advantage to the host. Thus, even some DNA we used to call "junk" is useful.

THE LANGUAGE OF LIFE DIFFERS VERY LITTLE BETWEEN INDIVIDUALS

With the ability to obtain DNA sequence information readily from multiple individuals, we are learning that our instruction books are remarkably similar, regardless of where our ancestors came from. If a sequence of my DNA is lined up with DNA derived from someone in Europe, Africa, or Asia, on average there would only be four differences between us in every 1,000 DNA letters. If we examined the entire genome, the differences between my DNA and that of someone from Asia or Africa would be slightly greater than those between me and a European neighbor, but 90 percent of the tiny amount of variability in my genome would exist between me and other Europeans, and only 10 percent would show some geographic significance with other populations. This remarkable similarity of humans across the world is a reflection of our shared historical origin. Population geneticists, looking at these data, are convinced that all humans are descended from a common set of about 10,000 founders, who lived in East Africa about 100,000 years ago. Thus we are truly all one family, and it's no wonder that our "DNA dialects" are remarkably similar.

As a first approximation, your genome is the same in every one of the 400 trillion cells of your body. But different cells use a different suite of genes to carry out their functions, and that's what makes a liver cell different from a brain cell or a muscle cell. These different programs are carried out by various proteins that bind to the DNA, and switch on or off the genes nearby.

Each time a cell divides, the entire genome has to be copied. But mistakes can creep in. Occasionally these can cause a cell to grow more rapidly than it should, and that can even lead to cancer. The environment can play a role here, in that carcinogens such as radiation and cigarette smoke increase the mistake rate of DNA copying.

BEYOND THE HUMAN GENOME PROJECT

Since 2003, rapid progress has been made by building upon the foundation provided by the original sequence of the human genome. Much more hard work is needed, since our ability to interpret the more than 3 billion letters of our own language of life is still rudimentary, and requires many other sources of information in order to make sense of this vast sea of data. With the cost of *DNA sequencing* continuing to drop at a breathtaking rate, it has become possible to determine the complete genome sequence of a vast array of other organisms, including hundreds of microbes and dozens of invertebrates and vertebrates. For some of these organisms, such as mice, rats, and dogs, having the genome sequence is valuable in its own right, since substantial research communities are devoted to understanding their biology. But all of these sequences inform us about the human genome, too. After all, if a particular segment of human DNA shows strong connections with other mammals, or even with species that lie farther away in the evolutionary tree, this particular segment must have performed an important function that does not tolerate much variation over evolutionary time.

Comparing Genomes Is Like Cryptography

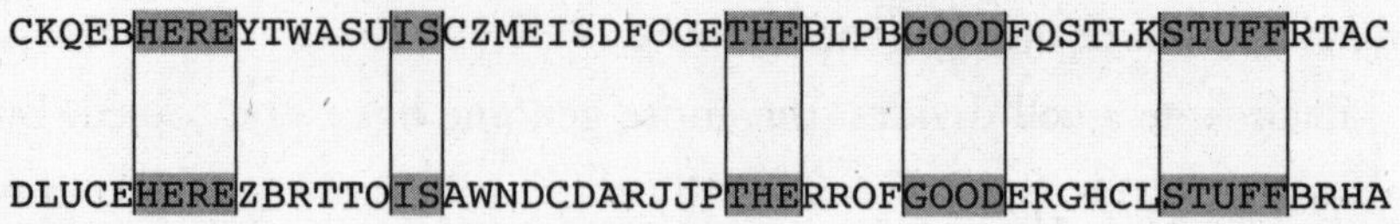

CKQEBHEREYTWASUISCZMEISDFOGETHEBLPBGOODFQSTLKSTUFFRTAC

DLUCEHEREZBRTTOISAWNDCDARJJPTHERROFGOODERGHCLSTUFFBRHA

Figure 1.4: Comparing genomes from different organisms is a powerful method to identify the parts that are most functionally important, as these will have been the most constrained during the process of evolution.

Figure 1.4 shows in simplified form how this kind of comparison can reveal genome features that may otherwise be hidden in a sea of gibberish.

Remarkable strides have also been made in characterizing human genetic variation. That of course is critical—if we hope to discover the hereditary factors that influence virtually all diseases, we need a complete understanding of the 0.4 percent of the genome that differs between individuals.

As the cost of DNA sequencing has continued to plummet, the potential of determining the complete instruction book of individuals for medical purposes has become increasingly more realistic. Only five years after the completion of the first human genome sequence, a project has been mounted to sequence 1,000 or more human genomes, drawn from individuals from all over the world. This will provide the most detailed view yet of genetic variation.

Many other large-scale research projects are now aimed more directly at determining genome function. The Encyclopedia of DNA Elements (ENCODE) is a project that involves dozens of laboratories working together to identify all the functional elements of the genome (the "parts list") and to determine how those work together to turn genes on or off in particular tissues.

Other projects are studying model organisms, including a project

that aims to inactivate (or knock out) each of the genes in the laboratory mouse. Since more than 95 percent of mouse genes have clear
matches to genes in the human genome, this powerful resource will
help determine the function of thousands of mouse and human genes,
one at a time.

The consequence of all this progress is that a new science has appeared at the very center of biology and medicine: you could call it
DNA cryptography. We've intercepted a highly elaborate message of
critical importance for the future of the human species. It is written
in a strange and seemingly impenetrable code, disarmingly simple in
its use of just four letters, but complex enough that decades will be
required before a combination of human ingenuity, laboratory investigation, and elaborate analysis on the most powerful supercomputers
will reveal the full secrets of the code. But what an amazing adventure
this is!

HOW DOES ALL THIS RELATE TO PERSONALIZED MEDICINE?

We are all individuals; in matters of health and disease, we bring
our own genomes, our environmental exposures, and our choices to
the table. And most of us live and make those choices with plenty of
mixed messages and motivations. We know we should exercise regularly, and eat healthy foods, but we don't always manage to do so. We
have information about risks and health all around us, but still we
sometimes throw caution to the winds. Young people especially live
for the moment and worry little about the future. It's older people,
starting especially at parenthood, who tend to be more circumspect.

Because you have picked up this book, I assume that you are interested in learning how to improve your chances of staying healthy.

What if I told you that the single most important source of information about your future health, and your risk of illness, and that of your parents and your children, was readily available, provided a window into your genome, was free, and required only about an hour or so to collect? Most of us ignore this powerful tool, even as we take care to fasten our seat belts, avoid potentially dangerous food, and try to make time for exercise. It's our family health history.

Every medical student is taught to record a family history as part of the evaluation of a new patient, but the actual purpose of the history is not always made clear, and all too often the process is rushed and cursory. How many of our medical records contain the utterly useless notation "noncontributory" in the section on family history? (You may never have even seen your records. Trust me; it's probably there.) This is a truly wasted opportunity.

Family health history turns out to be the strongest of all currently measurable risk factors for many common conditions, incorporating as it does information about both heredity and shared environment. Having a parent or sibling with cardiovascular disease doubles your risk. Having two or more of these "first-degree relatives" with heart disease, if they developed the disease before age 55, multiplies your risk fivefold.

Having a first-degree relative with colon, prostate, or breast cancer increases your risk two- to threefold. Similar risks are found if you have close relatives with diabetes, asthma, and osteoporosis. Surely this is the kind of information you and your doctor should know and incorporate into your own health care, in precise detail. Yet all too often the information is not collected, or is simply ignored.

In an attempt to rectify this situation, my colleagues and I joined with the then United States surgeon general Dr. Richard Carmona in starting a family health history initiative in 2004. This Web-based resource (http://familyhistory.hhs.gov) makes it easy for people to collect their own family health histories conveniently, in their own homes. It helps encourage people to call or e-mail relatives to obtain missing information. With the use of a Web tool that is privacy-protected and freely available from the U.S. Department of Health and Human Services, each family health history can be entered into a standard form, which produces the kind of "pedigree" that health care providers need, and can be readily integrated into the electronic health record.

Hundreds of thousands of individuals have taken advantage of this opportunity, and that number is growing daily. (At the end of this chapter you can find detailed instructions about how to carry out this process for yourself and your family, and I strongly urge you to do this.)

It is profoundly unfortunate that our medical care system has largely failed to encourage this kind of data collection. A recent survey by the Centers for Disease Control and Prevention (CDC) suggests that less than 30 percent of Americans have actively collected health information from our relatives, though 96 percent believe this information is important.

Of course, family health history has limitations. Many individuals are stricken by common diseases such as cancer, diabetes, heart attack, or Alzheimer's disease despite an absence of any relevant family history. And, of course, adopted individuals often lack access to this information.

THE NEW PARADIGM

The revolution that now promises to transform our physical *and* mental lives is the opportunity to combine this knowledge of family history with a survey of your entire DNA instruction book, and identify the specific glitches hiding in your life script. And let's be clear about this: we all have these glitches. If you started this book with the notion that you are a perfect genetic specimen, I have bad news for you. There are none. Within that 0.4 percent of variation that distinguishes you from other members of our species are lots of bits that probably have no significance for health. But there are some that place you at risk for future illness. We are all flawed mutants.

How quickly will this affect your health care? Most change that affects us profoundly does not occur overnight but emerges gradually over a period of time. Then one day we look around and realize that the world is a different place. When I wrote my first e-mail message using a clunky software system in 1989, I never dreamed that it would become my primary means of communication with colleagues, friends, and family. When the Human Genome Project was begun, no one imagined that it would utterly transform the way questions in biomedicine were posed and answered. The ancient, traditional medical paradigm is now shifting so gradually that you may not yet have noticed it. But the consequences will be profound.

In the old way of thinking, a diagnosis of disease was based upon the presence of certain symptoms, supplemented by various laboratory tests. Our treatment was based upon studies that had been done on hundreds or thousands of people with the same diagnosis, who were all considered to be essentially identical subjects. On the basis of this paradigm, we have spent $2 trillion a year in the United States alone on health care. Only a tiny fraction of that is devoted to prevention—the main focus is on disease. We do not have a health care system; we have a sick care system! The treatments we use have largely been arrived at by trial and error, often without a clear understanding of why they work—or why they often don't.

The new paradigm is radically different. We know that each individual is unique, endowed with certain genetic variants that may provide advantages, and others that cause vulnerabilities to future disease. While some of those vulnerabilities are highly predictive of future trouble, most of them involve considerably milder risk, and will give rise to illness only if combined with other genetic risks and various environmental triggers. Disease is not random, nor is it unavoidable. Personal choices have a profound impact on your health, and they are your responsibility (not just your doctor's).

If we're not careful, the increasingly accurate predictive power of personalized medicine could even begin to blur the concept of a diagnosis. Is an individual with a 60 percent risk of colon cancer already ill? Does my 35 percent chance of glaucoma count as having the disease? No and no. We must seriously guard against that kind of slippage in semantics. Diagnoses should still be reserved for those who have developed actual symptoms of illness. But the precise description of individual illness is likely to be greatly enhanced by specific molecular information. Diseases that we used to lump together under one label will be split apart into separate conditions with different prognoses and different therapies. In other instances, diseases that we used to think were utterly unrelated will turn out to have a common pathway and to benefit from a common therapeutic approach. For instance, drugs developed for cancer may turn out to work for arthritis or Alzheimer's disease.

PERSONALIZED MEDICINE IS ALREADY HERE

If all of this sounds like science fiction, that is certainly not the case for the millions of people whose lives have already been significantly affected by the revolution in personalized medicine. Let's consider a real-life example. Of all the conditions where DNA analysis has already had an impact, what could be more dramatic than sudden death?

The phone call Doris Goldman received on that morning in 1979 was a mother's worst nightmare. Her 20-year-old son, Jack, a handsome, athletic college student, had been found dead in his sleeping bag in Wyoming. Extensive postmortem examination, including studies of all possible drugs and toxins, revealed absolutely no cause. Two years later Doris's daughter Sharon, just 19 years old, suffered a cardiac arrest. Though she was resuscitated, she had significant brain damage and struggled mightily over the next few years to recover. She went on to attend community college, get married, and have a son. But then the unspeakable nightmare recurred: Sharon was found dead one morning at the age of 29. Again, no cause could be identified.

Some mothers in this situation might have sunk into depression, anger, or blame. Not Doris. She collected medical histories and electrocardiograms (EKGs) from her large extended family. She learned of another cousin who had died at age 45 in her sleep. A review of the EKGs by cardiologists Doris had recruited began to reveal a possible answer. Specifically, a component of the electrical conduction pattern that the EKG detects, known as the QT interval, was prolonged in quite a number of family members.

This condition, known as "long QT syndrome," had been described in a few families in the medical literature and was, in fact, associated with fainting spells and sudden death, as it predisposed its victims to a potentially fatal heart rhythm called ventricular fibrillation. Careful review of EKGs that had been done in the past on Sharon, read as normal at the time, revealed subtle but convincing evidence that she too had this condition.

The differences between a normal QT interval and one that could be dangerous are quite small, and so in Doris's family it was not entirely possible to identify who was at risk by reviewing the EKGs. But in 1996, tools arising from the Human Genome Project made it possible for specific genes for long QT syndrome to be identified. Doris's family turned out to have a mutation in a gene, *HERG*, which is normally involved in sodium transport across the cardiac muscle cell membrane. With a specific genetic test now available, no fewer than 37 members of Doris's family were found to have this mutation, and to be at risk for the same sudden death as Jack and Sharon. Though she herself had never had a blackout spell, Doris found that she, too, was on the list of mutation carriers, as were her surviving daughter and Sharon's young son.

This sounds like a grim scenario, but it was not a hopeless one. In this instance the value of the information was profound. Research studies have convincingly demonstrated that individuals with mutations linked to long QT syndrome can have their risk greatly reduced by lifelong treatment with a class of cardiac drugs called beta-blockers. Members of Doris's extended pedigree have now been treated, and there have been no further deaths. All of the affected family members also have automatic external defibrillators in their homes (if they can afford this device), and they make sure that family members are trained to perform resuscitation if the need should arise. Some have even had automatic defibrillators implanted in their chest.

Long QT syndrome is not a disorder that most members of the public, or even most health professionals, have heard of. Yet it turns out to be a critically important condition to recognize. The availability of DNA testing has made it clear that as many as one in 4,000 individuals in the U.S. may be at risk. With some families, death seems to occur during sleep. In others, death strikes at times of exertion or strong emotion. Perhaps the most dramatic example of this is the story of a family who lost two sisters on the same day. It was Super Bowl Sunday. One sister was shoveling snow around her home in Virginia. She suddenly fell dead. As word spread through the family, a second sister, distraught at the news, collapsed and could not be resuscitated despite the efforts of emergency personnel. It turned out that these two sisters, and six of their siblings, carried mutations in one of the genes for long QT syndrome, as did a number of their children. The tragedy on Super Bowl Sunday provided this family with a chance to find out about this potentially devastating condition, and probably saved the lives of many others, even as they continue to grieve for their lost loved ones.

Several lessons can be derived from the dramatic stories of these families. First of all, knowing your family history can save your life. In the midst of tragedy, investigating the causes of unexplained deaths has led to a chance at a full life for relatives who might otherwise have faced a similar fate.

Second, health professionals don't always know the answers. In these families with long QT syndrome, the sudden death of a young person did not immediately set off the right lightbulb in the minds of the physicians. Motivated individuals in families can make all the difference.

Third, DNA testing, although not always as clear-cut as in these cases, can in the proper circumstances provide much-needed answers, and can even provide powerful predictions of risk for other family members.

Fourth, even for long QT syndrome, a highly heritable condition, it is clear that the environment has a strong role, since cardiac arrest is most likely to occur in particular settings. Another critical environmental influence on long QT syndrome comes from over-the-counter or prescription drugs, many of which can increase the likelihood of fainting or sudden death and must be avoided by individuals with this condition.

Fifth, none of us should ever be fatalistic about a serious condition, even though it may be written into the DNA of every one of our cells. We will not be in a position to alter our own genomes for a very long time to come, but other medical interventions can have profound benefit.

Finally, one more point should be made about this condition and its broader implications for the rest of us. Although, as noted above, only about one in 4,000 individuals is affected by the long QT syndrome in this highly heritable way, studies of hundreds of individuals have shown that there is considerable variability in the length of the QT interval among otherwise normal individuals. Furthermore, those in the upper end of this distribution face about a threefold increased risk of sudden death, even though they do not actually have long QT syndrome. Recently, variations in several genes have been identified as playing a role in "normal" variations in the QT interval. Though measurement of the QT interval or the genes contributing to it in otherwise normal individuals has not yet emerged as part of personalized medicine, this would not be an unreasonable addition to the information to be collected on individuals in the future, especially given the potential for prevention.

Few individuals have heard of long QT syndrome. Fewer still have encountered this diagnosis in a family member, or have undergone a DNA test for the condition. But as we shall see, not all genetic conditions are rare. If you have children or grandchildren who are less than 35 years old, chances are that they have had a genetic test. If you are a woman with children under 30, there is a fair chance you have had one yourself, although you may not have been fully aware of it. In many ways, personalized medicine is already here.

WHAT YOU CAN DO NOW TO JOIN THE
PERSONALIZED MEDICINE REVOLUTION

Take advantage of the U.S. surgeon general's Family Health History Initiative and the tool "My Family Health Portrait." Go to http://familyhistory.hhs.gov/ and learn how you can collect medical information from your family to construct a standard medical pedigree. Once you have put this all together, send copies to all family members. Take your own copy to your next visit with your health care provider, and use this as a means to start a conversation about what your own personal risks for future illness might be, and what you can do about them.

CHAPTER TWO

When Genes Go Wrong, It Gets Personal

On a chilly California morning in 1972, a young Japanese woman went into labor. Her husband, a German physicist, was away at a meeting, so she drove herself to the hospital.

What could be more normal and more blessed than the opportunity for this young couple to bring new life into the world? But this birth was destined to be anything but usual. The first surprise came when the examining physician heard two heartbeats. Shortly afterward, the *identical twins* Anabel and Isabel arrived on the scene—to all appearances two normal, healthy baby girls.

But the second and much more ominous surprise became apparent just three days later, when Anabel showed dangerous signs of intestinal blockage, and emergency surgery was required to save her life. Recognizing that this could be a complication of cystic fibrosis (CF), the attending physician ordered sweat tests to be conducted on both girls. The results were unequivocal and devastating. These identical twins, who carried identical DNA, were both affected with this genetic disease and faced a very serious threat of a shortened life span. In fact, the physicians told the young couple that they were in for great struggle and sorrow, as it was not expected that either twin would live to see her tenth birthday.

Cystic fibrosis is the most common potentially fatal genetic disease in individuals of northern European background, but it is quite rare in Japan. The twins' father, being mathematically inclined, calculated the combined odds of the occurrence of cystic fibrosis in identical twins born to a German and Japanese couple as one in 1.8 billion. But rare genetic diseases do strike real people, and in this case, the probability of cystic fibrosis in Anabel and Isabel was 100 percent.

I met Anabel and Isabel in 2008. At age 35, they both looked remarkably well. Both of them had graduated from Stanford University. Anabel is a genetic counselor, and Isabel is a social worker. Their gritty story of struggle, determination, and an unwavering commitment to beat the odds, including double lung transplants for both of them, is described in their moving dual autobiography, *The Power of Two: A Twin Triumph Over Cystic Fibrosis.*

DOMINANT, RECESSIVE, AND ALL THAT

Diseases like CF, sickle-cell anemia, or Huntington's disease are the predictable result of mutations in a specific gene, and are referred to as "single-gene," or Mendelian, diseases. These are the simplest to understand at the DNA level, and discovering the causes of hundreds of these conditions represented the first wave of the genomic revolution. To understand this class of disease, we need to consider some of the basics about genetic inheritance. The good news about genetics is that a few basic principles will allow you to figure out answers to many different situations. This branch of science is blessedly free of the need to memorize mountains of factual material. Perhaps that is why scientists like me, who prefer simple principles and hate memorizing stuff, migrate into this field, instead of going into neuroscience or immunology. But let's try to get a few of these principles clearly spelled out, as a bit of "Genetics 101" will inform all the future information about personalized medicine in this book. (More details are in Appendix B.)

Principle 1: We humans are *diploid*. That means each of us carries two copies of almost all the genes in our instruction book, one inherited from our mother and one from our father. Genes are carried on *chromosomes*, which can actually be seen under a microscope when the cell is about to divide. Figure 2.1 shows human chromosomes from a normal male, arranged in order to demonstrate the pairs. It is apparent that chromosomes come in different sizes and different banding patterns, but they are all paired except for the *X chromosome* and the *Y chromosome* in a male. A female has two X chromosomes instead.

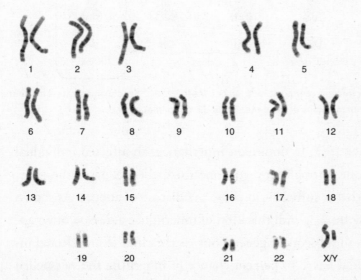

Figure 2.1: The chromosomes of a single cell from a normal human male. A female would have two X chromosomes instead of an X and a Y.

Principle 2: In a recessive disease like cystic fibrosis, *both* copies of the responsible gene must contain misspellings for the disease to occur. As shown in Figure 2.2, that can happen only if each parent carries one misspelled copy and passes it on to the child. The parents in this situation are known as *carriers*, and in the case of a recessive disease they are generally completely normal and unaware of their status. Each child of carrier parents has a one-in-four chance of being affected. I discovered from DNA testing that I am a carrier for alpha-1-antitrypsin deficiency and for hemochromatosis. But neither has affected my own health.

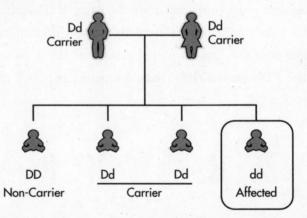

Figure 2.2: Recessive inheritance, as occurs in cystic fibrosis and sickle-cell anemia. "D" is the normal copy of the gene, "d" is the abnormal copy.

Principle 3: In dominant inheritance, an affected individual has one normal copy and one misspelled copy of the gene, and that is sufficient to cause the disease to appear. As shown in Figure 2.3, with this kind of inheritance a disease often appears in subsequent generations, as the child of an affected individual has a 50 percent chance of inheriting the misspelled gene, and also being affected. Well-known examples of dominant genetic diseases include Huntington's disease and neurofibromatosis (sometimes erroneously referred to as "Elephant Man disease"—the Elephant Man actually had a different condition). Another example of a dominant condition is long QT syndrome, referred to in Chapter 1.

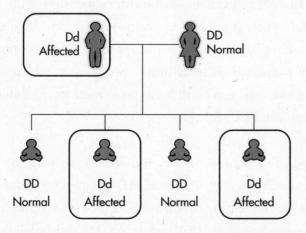

Figure 2.3: Dominant inheritance, as occurs in Huntington's disease.

Principle 4: The inheritance of most genetic conditions is not this simple. Most misspelled genes cause a *predisposition* to, but not a *predetermination* of, disease. Geneticists sometimes refer to this as "incomplete penetrance." Simply put, that means someone who carries a particular gene capable of conferring a risk of disease does not always experience the consequences. The *BRCA1* gene in my own family is an example of incomplete penetrance. Specifically, women who carry a *BRCA1* mutation have about an 80 percent lifetime risk of developing breast cancer and a 50 percent risk of developing ovarian cancer. But this means that some women with these mutations never develop cancer at all. The penetrance is even lower in males, who despite *BRCA1* mutations face only a modest risk of cancers of the pancreas, prostate, and male breast.

Principle 5: Although virtually all common diseases—such as diabetes, heart disease, and cancer—have hereditary components, there are multiple genetic risk factors that contribute to these conditions. We call these diseases *polygenic*. The power of each individual genetic risk factor is generally quite low, and so illness is likely to occur only with a combination of several of them, along with appropriate environmental stimuli. I reported on some of my own genetic risk factors in the introduction, and I will have much more to say about this situation in subsequent chapters.

But now let's return to Anabel and Isabel, and focus on cystic fibrosis as a cardinal example of a disorder caused by mutations in a single gene that is yielding up many of its secrets as a consequence of the genome revolution. In 1972, when Anabel and Isabel were born, not much was known about cystic fibrosis, other than its recessive inheritance and the fact that it affected numerous organ systems of the body. It was known to involve the pancreas (the name "cystic fibrosis" refers to the formation of cysts and fibrous scars in the pancreas of affected individuals), resulting in an inability to secrete digestive enzymes. If these enzymes were not added to the person's diet, profound malnutrition would result. The intestines were known to be involved in some cases—as with Anabel, who required emergency surgery shortly after birth for a serious blockage. Males with CF who lived to adulthood were also noted to be infertile. Most significantly, however, the lungs were known to be seriously affected. Thick, sticky secretions accumulated, followed by recurrent infections, destruction of lung tissue, and all too often an early death.

Years ago, mothers of children with CF noted that when the children were kissed, their skin tasted salty. As a result, a somewhat bizarre diagnostic test was developed for cystic fibrosis: the measurement of chloride levels in sweat. Salty sweat suggested that there might be some problem in the transport of salt and water, and that this same problem might perhaps affect the lungs, the intestines, and the pancreatic ducts, but it was not until the 1980s that a definite connection was demonstrated. Even then the information was insufficient to identify the responsible gene.

My laboratory played a central role in the identification of the genetic mutation in CF. But it took many long years of torturous work, because of the lack of information about the human genome at that time. Families with several affected children were asked to participate in research, in order to try to map the gene to a particular location in the genome. The principle of the method was simple. Since this is a recessive disease, affected siblings must share identical DNA containing the CF gene on both their maternal and their paternal chromosomes, whereas elsewhere in the genome they can be expected to share only 50 percent of their DNA. From studying a very large number of such families, it ultimately became clear that a long stretch of DNA on chromosome 7 must contain the CF gene. But the remaining task was daunting: the region of DNA involved was approximately 2 million base pairs, and the methods for dealing with such large stretches of DNA in the 1980s were very slow and imperfect.

Teaming up in 1987 with an investigator from Toronto, Dr. Lap-Chee Tsui, my lab trolled through this large, uncharted DNA territory, searching for any subtle mutation that might distinguish CF patients from unaffected individuals. After many false starts, and many occasions when the hopes of our research teams were dashed by the next day's data, the answer finally emerged. I can recall the exact moment when we were sure we had it. Lap-Chee and I were attending a meeting at Yale University, and had set up a fax machine in his room to monitor the work in our labs during that week. Rushing back to the room after that day's sessions, we saw the evidence that a simple deletion of three letters (CTT) of the genetic code, located in the middle of a gene of unknown function, appeared unequivocally to track with the presence of CF.

The genes for other inherited conditions had been identified earlier, with the use of functional information about the disease, or thanks to rare patients with visible chromosome rearrangements that pointed directly to the responsible gene. But this was the first time that the cause of a human genetic disease had been revealed without the benefit of any of those clues. The identification of the CF gene set the stage for the subsequent unveiling of the genetic causes of nearly 2,000 conditions over the next 15 years.

Figure 2.4 shows the nature of the mutation that we discovered in the DNA. What is shown is only a very small part of a rather large gene, *CFTR*, which codes for a protein of some 1,460 amino acids. The deletion of three letters in the DNA sequence results in the loss of a single amino acid (phenylalanine) from this large protein (number 508 of the 1,460). This is indicated by the notation ΔF508. The deletion of just three letters out of 3 billion, falling in this particular place, results in a multi-organ disease that affects approximately one in 3,000 individuals of northern European background, causing untold distress and difficulty for these individuals and their families. Even after having worked in the field of genetics for almost 30 years, I am still amazed to contemplate how such a subtle change can have such dramatic effects.

We published the information about the CF gene in September 1989. On the cover of that issue of *Science* magazine was a photograph of a five-year-old boy with CF, Danny Bessette. I recently ran into Danny at a reception. I was relieved to see that, like Anabel and Isabel, he is doing well despite many significant medical challenges.

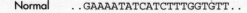

Normal   ..GAAAATATCATCTTTGGTGTT..

ΔF508   ..GAAAATATCAT---TGGTGTT..

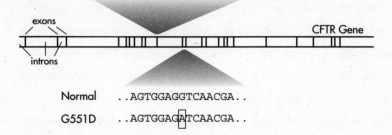

exons

CFTR Gene

introns

Normal   ..AGTGGAGGTCAACGA..

G551D   ..AGTGGAGATCAACGA..

Figure 2.4: A diagram of the *CFTR* gene. This gene normally codes for a protein that transports salt and water across cell membranes in various organs. But if both copies of the gene are misspelled with mutations such as ΔF508 (a deletion of CTT) or G551D (a substitution of A for G), the result is cystic fibrosis.

In the past two decades, many more details have emerged. The original three-base deletion we discovered remains the most common cause of CF, but it is a theme with variations, as now more than 1,000 mutations have been identified worldwide that can cause CF. Geneticists call these different mutations of the same gene *alleles*, though this is among the genetic terms that are unfortunately obscure to most nonscientists; it is also difficult to pronounce, and sure to be a conversation-stopper. For the purposes of this book, I will occasionally use it, but I will also use two more familiar terms: *mutation*, for DNA misspellings that have negative consequences; and *variant*, for spelling differences of all sorts, negative, positive, or neutral. Mutations are bad variants. Many variants are just fine, the spice of life. Some are even helpful (this is how evolution works).

Not all mutations in *CFTR* are equally awful. Some less common mutations seem to spare the pancreas. Others cause infertility but nothing else.

Even in cases of single-gene recessive diseases like CF, the course of the disease can vary in individuals with exactly the same mutations. Individuals who have two identical copies of the mutation we discovered (about half of CF patients) can differ widely in the severity of their lung disease. Why should that be?

One contributor to this variation turns out to be other genes in the genome that serve as "modifiers." Most genes have some degree of normal variation. Those normal variations in other pathways can play a role in affecting the severity of a genetic disease like CF. Already several such modifiers have been identified for CF.

Another important modifier of severity is the environment. A recent study demonstrated that exposure to secondhand smoke can play a significant role in the progression of CF lung disease. And of course the dramatic improvement in survival in CF (Figure 2.5) cannot be due to changes in the gene pool in a period of just 50 years. In this case, the development of better medical interventions has provided beneficial environmental influences on the disease. These include the availability of pancreatic enzyme capsules to improve nutrition, the use of vigorous chest physical therapy to clear the sticky secretions that otherwise lead to lung infections, the use of aggressive antibiotics to keep infections at bay, the development of an aerosolized enzyme therapy that can digest the sticky DNA in the lung secretions and make them easier to clear, and the use of saltwater mists to assist in keeping the airways clean.

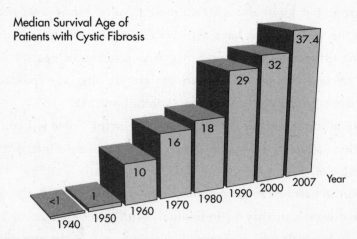

Figure 2.5: Medical research has led to dramatic improvement in survival for individuals with cystic fibrosis.

Most dramatically, when all these efforts fail, double lung transplantation has been lifesaving for hundreds of CF patients such as Anabel and Isabel, though the availability of organs continues to be a major challenge, and the risks of rejection are substantial. In fact, Anabel has already gone through an episode of rejection and has undergone a second lung transplant.

FROM GENE TO CURE?

It is one thing to learn the specific mutation behind a disease. It is quite another to overcome it. Many people scoff that therapy for rare genetic diseases will never be practical. Consider sickle-cell anemia. This was the first recessive genetic disease identified. It is common among individuals whose ancestors lived in areas where malaria has historically been widespread: around the Mediterranean, in Africa, and in Southeast Asia. Sickle-cell carriers—those with one, not two, copies of the sickle mutation—are better able to survive childhood malaria than those who do not have this mutation. Those individuals carrying two copies of the sickle mutation, however, have a serious blood disorder associated with frequent painful crises and a serious limitation in life span.

The genetic mutation in sickle-cell disease, located in one of the genes for hemoglobin, has been known for 50 years, yet this information has led to relatively little in the way of novel therapeutic approaches. So why should I claim that genetic medicine is at a tipping point now? For one thing, the pace of medical progress is not linear. Fifty years of slow progress does not imply that the next 50 (or even the next 10) will be similarly slow. Most researchers actually predict substantial advances in the treatment of sickle-cell anemia in the coming decade. A major reason for their optimism is the promise of gene therapy, though there have been many ups and downs in this field. To apply this approach to sickle-cell anemia, it will be necessary to insert a normal copy of the relevant hemoglobin gene into the bone marrow of an affected person and get it to function efficiently over time. Progress with other diseases provides encouragement that this approach may eventually work for sickle-cell disease. And gene therapy is not the only option for breakthrough treatments—new ideas are also emerging about the development of drugs that could block the sickling of red blood cells in affected individuals.

What about CF? It has now been 20 years since the discovery of the cause. What strategies are being pursued to use this information therapeutically, and is there any progress to report?

Immediately following identification of the *CFTR* gene, there was much enthusiasm for a gene therapy to cure this disease. The idea was relatively straightforward: if one could stitch a normal copy of the gene into a virus that likes to infect the airways, then patients could "breathe in a cold," so to speak, and the "cold" could implant copies of the healthy gene. We demonstrated this principle in a laboratory culture dish within a year after finding the gene, showing that such a virus could correct the problem with salt transport in cultured airway cells. But the challenge of efficient gene delivery in living individuals has proved to be extremely difficult. Obstacles arise in several ways. First of all, the efficiency of delivery has to be very high, because correcting just a few cells in the airway will have no benefit. The virus preparation therefore has to work like a large army, spreading out and occupying a lot of territory. Furthermore, the normal copy of the gene not only must be ferried into the cell by the virus, but must then be retained in a stable fashion and expressed in reasonable quantities as RNA and protein.

Finally, all this must be accomplished in a way that evades detection by the immune system, or the body's own defenses against this viral infection may rapidly eliminate any benefit.

Unfortunately, all these issues have presented roadblocks. To understand the situation a little better, consider a sports analogy. You can think of your 20,000 genes as a well-organized athletic team, working together to try to win the game of life. A CF mutation represents a player who is seriously injured and has to be carried off the field. Gene therapy—the game strategy—tries to compensate by sending out another player, but that player has to be capable of finding his way to the appropriate place on the field, and then performing well without being injured himself. If he subsequently collapses and is carried off himself (this is similar to the situation of a gene that gets into the cell but functions for only a short time), then the problem is not solved. And if a substitute player turns out not to be on the official team roster, the referee (the immune system) will call a halt to the game and throw the new player out.

Given all these challenges, the progress of gene therapy over the course of the last 25 years has been frustratingly slow. A few years ago, success finally seemed to be at hand, when a small number of children with a rare form of immune deficiency caused by a defective gene were apparently cured by a therapy that supplied them with the missing gene, carried on an inactivated virus. But the celebrations turned somber a few years later, when several of these children developed leukemia, apparently as a result of the inadvertent activation of a cancer gene by that same virus. I will have more to say about gene therapy in Chapter 10.

Efforts to develop gene therapy for CF continue, but the early hope for a quick answer has been replaced by the realization that this will take many more years of hard work. Meanwhile, another approach, empowered by the discovery of the gene, is showing considerable promise. Coming back to our athletic team, wouldn't a good solution be to quickly treat the injuries of the original player, bring him back to full health, and send him back on the field? Though our analogy is getting a bit stretched, that is in effect what a drug therapy aims to do.

In the past most drugs were developed by largely empirical methods. The compounds that were explored as possible therapies were often natural products derived from bacteria, fungi, or plants, and only a few could be tested. But a new and more comprehensive "designer drug" approach that defines the specific target for the drug, and then screens hundreds of thousands of candidate drugs to find the winners, has begun to replace those older, less systematic methods. For CF, knowing the precise molecular defect has made it possible to identify candidate drugs that act to correct the salt-transport defect in cultured airway cells. Once it has been shown that these compounds are nontoxic in animals, clinical trials can begin. More information about how drugs are developed can be found in Appendix D.

The results of an early trial of one of these designer drugs on three dozen individuals with CF have been truly exciting. One of them is Bill Elder. Bill is not only a young man with CF, but also a student at Stanford, working in the laboratory of Dr. Jeffrey Wine, whose interest in CF was sparked when his own child was diagnosed with the disease. Bill has one of the less common, "sick player on the field" variants of the disease, denoted G551D (Figure 2.4). Prior to the new drug trial, he had been doing relatively well, although he required multiple medications and daily physical chest therapy, as well as occasional intense antibiotic treatments.

As a volunteer in the trial of the CF "potentiator," known as VX–770, Bill had to take just three white pills twice a day, and then undergo numerous tests to see whether the drug was working.

The results of the initial trial were nothing short of astounding. The sweat chloride levels for treated individuals dropped to nearly normal. One particular test of salt transport carried out by measuring tissues in the nose had nearly perfect results. Most dramatically, the air flow in the lungs improved over just a two-week period. No side effects of the drug were noted. Although this is only a brief, early test of the therapy, no previous attempts at drug treatment of CF had accomplished such milestones.

Though much work remains to be done, and it would clearly be premature to declare victory over CF, these developments provide the brightest ray of hope for this condition in many years.

A decade ago, speaking to the annual North American Cystic Fibrosis Conference, a gathering of caregivers, patients, and families, I concluded my presentation by asking the group to join me in a song of hope for the future. Few eyes were dry as thousands of people rose to their feet and sang along with the chorus:

> *Dare to dream, dare to dream,*
> *All our brothers and sisters breathing free.*
> *Unafraid, our hopes unswayed,*
> *'Til the story of CF is history.*

That dream seems closer than ever now. Moreover, these advances for CF represent the first wave of a growing ocean of potential new treatments for diseases that are flowing in from the world's laboratories, thanks to our new ability to read the secrets of the language of life.

YOUR DIET CAN SAVE YOUR LIFE

Another therapeutic avenue, distinct from genes and drug therapies, is in other environmental approaches to treating disease. An important part of your environment is the food you eat. Now that we are learning the precise mechanisms of many diseases, in some instances we can precisely counter their effects just by modifying the diet.

Tracy Beck is a 35-year-old PhD in astrophysics, working at the Space Telescope Science Institute, developing the next version of the Hubble Telescope. But if Tracy had been born just 10 years earlier, she might now be institutionalized with severe mental retardation, seizures, and a small and underdeveloped brain—that is, if she was even still alive.

Tracy appeared completely normal at birth, but during her first month her mother was troubled by her sleepiness, especially in comparison with her older sister. A newborn screening test revealed that her blood phenylalanine level was nearly 10 times normal. Phenylalanine is an essential amino acid present in all forms of protein. Tracy has a genetic glitch that robs her of an enzyme, phenylalanine hydroxylase, necessary to properly metabolize this substance, so she has far too much of a good thing. Even though this amino acid is essential for life, such high levels cause severe toxicity in the developing brain.

Tracy's parents were shocked by the diagnosis but immediately set about instituting the rigorous dietary treatment that had been developed for this rare condition. Specifically, individuals with this disorder, called phenylketonuria (PKU), have to follow a lifelong diet of extremely limited protein intake (to keep phenylalanine low), supplemented by a formula that provides all the other amino acids in quantities necessary for growth. You can imagine the challenges of keeping a child on such a rigorous diet, amid school lunches, birthday parties, and sleepovers. Tracy admits that at age nine she rebelled against the dietary restrictions, and began sneaking forbidden foods, especially cheese. Over a few months, her outstanding class performance deteriorated, and she was even briefly placed in a remedial math class. Realizing that the consequences of her rebellion might be severe in her future, Tracy and her parents arrived at a diet that she could tolerate and worked hard to educate others about the importance of maintaining it. To this day, she remains on this unusual form of dietary management, simply telling friends, "I have a medical disorder, and I have to avoid protein," when she is in social situations where others are enjoying high-protein foods.

It's an extreme diet, far beyond what people do to watch their cholesterol. For example, Tracy has to avoid any diet sodas that contain aspartame, as this artificial sweetener will be converted to phenylalanine in the body, and will cause dire consequences for someone with PKU. Despite these restrictions, Tracy has been a high achiever. She is one of the first individuals with PKU to attain a doctoral degree, and she serves as a wonderful role model for younger people with this condition. One of her major challenges has been persuading her health insurer to cover the cost of her special dietary formula, which runs to about $1,300 per month. One would think that a scientifically proven, highly effective therapeutic for an otherwise devastating disease would be a no-brainer for coverage by health insurance, but the dysfunctional system presently operative in the United States does not always respond to such compelling logic.

Since PKU is a recessive disease, by inference Tracy's parents must both be carriers. Both of Tracy's younger brothers also have PKU, each having been diagnosed within a few days of birth. This proves an important point about the mathematics of genetic risk. Although the risk to each child of carrier parents for a recessive disease is one in four (Figure 2.2), that risk has no memory, and thus a family of four siblings might have anywhere between zero and four affected children with the recessive disease. In Tracy's case, it's three out of four. (One sister escaped the others' fate.) Both of her brothers have also been well managed on the PKU diet, have graduated from college, and are pursuing careers in communications.

Phenylketonuria is the most compelling current example of a disease that is 100 percent genetic but whose consequences are 100 percent preventable by an environmental manipulation.

A second and very recent example of a dramatic advance in medical therapy for a genetic disease is exemplified by five-and-a-half-year-old Blake Althaus. Everyone commented shortly after his birth about how long and graceful Blake's fingers were, and a future as a piano virtuoso was imagined. His mother became concerned, however, when she noted a possible curvature of his spine. An ophthalmologist then noted an additional problem: dislocated lenses in the eyes. Of greatest concern was a cardiac ultrasound showing that the first part of the aorta—the largest artery in the body, coming directly out of the heart—was enlarged and would be prone to sudden tearing in the future, leading to a high risk of sudden death.

The parents were told that their son was affected with a particularly severe form of a condition known as Marfan syndrome. Given the rapid enlargement of the aorta, one physician predicted that Blake would probably not live past age two. In great distress, the parents searched the Internet and ultimately contacted Dr. Hal Dietz at Johns Hopkins University, a world expert on Marfan syndrome. Dr. Dietz reassured them that the prediction of such a short life was overly pessimistic, but warned that Blake would need close monitoring and might need surgery soon for the enlarging aorta.

But then something dramatic happened. Building on years of research, initially with laboratory studies of cells growing in a dish, and subsequently on a mouse model of Marfan syndrome, Dr. Dietz identified a drug therapy that might slow or stop the damage to the aorta. Better yet, the drug, losartan, had already been in use for more than 10 years for the treatment of hypertension, and was known to be safe in children.

And so, at the age of just 18 months, Blake was begun on losartan. His parents held their breath. Up until that time, successive ultrasounds showed Blake's aorta expanding steadily and dangerously. But a few months later, the enlargement had stopped. Over the next four years, Blake in effect grew into his aorta. Now that he is five and a half, this artery is almost within the normal range for a child of his age.

Marfan syndrome is caused by a mutation in the gene for fibrillin, an essential protein in connective tissue, including the aorta, the spine, and the fibers that hold the lens in place in the eye. When that mutation was discovered almost 20 years ago, most researchers assumed it would be extremely difficult to treat medically, since compensating for a defective structural protein was likely to be much more difficult than compensating for an enzyme that catalyzes a metabolic pathway. It's as if a brick house is built with bad bricks—you have to find and fix all of them. But Dr. Dietz and his team challenged this conventional wisdom, ultimately demonstrating that fibrillin has another important function: it binds to another protein, called TGF-beta. When fibrillin is defective, as in Marfan syndrome, TGF-beta circulates in abnormally high amounts. The researchers hypothesized that this internal overdose might contribute to the aortic enlargement. This was the motivation for trying losartan, since this particular blood pressure drug has an additional feature of serving as an antagonist of TGF-beta. In the early trials with severely affected individuals like Blake, the results were dramatic.

A large clinical trial is now under way to see whether losartan will benefit adults affected less severely than Blake. Famous adults who have had Marfan syndrome and died suddenly from a rupture of the aorta include Flo Hyman, the volleyball star, and Jonathan Larson, the author of the Broadway hit *Rent*. With the advent of losartan, it is likely that many of these tragic deaths can be prevented.

WHO WANTS TO SCREEN YOU, AND WHY?

There are many diseases such as cystic fibrosis or PKU, for which a particular biochemical or DNA test result makes a very strong prediction about the likelihood of illness, and interventions are available. Sometimes the test is done on the individual, in order to detect the presence of a genetic condition that may need intervention; sometimes it is done on prospective parents, to see whether they might carry a particular genetic mutation that, while no threat to their own health, poses risks to an affected child. In the new world of medicine we use the term *genetic screening* for tests applied broadly across the population, regardless of family history or prior medical history. Genetic testing means something more targeted, in a circumstance where there is felt to be an unusually high likelihood of a problem.

Newborn Screening

Tracy Beck, the PhD astrophysicist with PKU, represents an amazing success story of newborn screening. Most states in the United States began screening for PKU in the 1960s. Over time, more and more conditions have been added to the screening list; the focus has been on those for which screening can be technically successful, and for which early diagnosis is clearly beneficial. Sometimes these diagnoses can lead to institution of drug therapies; sometimes to special diets; sometimes to surgery or other options.

The March of Dimes currently recommends screening newborns for 29 conditions. About 4,000 babies each year are found to have one of these disorders. All states screen newborns for PKU, hypothyroidism, galactosemia, and sickle-cell anemia. The reasons are compelling. While hypothyroidism is generally not due to a single genetic cause, early detection is critical. If your child has hypothyroidism, it is essential to replace thyroid hormone immediately to allow normal brain development. Galactosemia is due to a mutation that prevents metabolism of the sugar galactose, which is present in milk and needs to be converted into glucose. This can be treated by a dietary restriction. Sickle-cell anemia (mentioned earlier) affects about one in 400 African-American babies. Early diagnosis can lead to vigilant medical care, including early treatment with vaccines and penicillin to reduce the risk of dangerous bacterial infections to which children with sickle-cell disease are particularly vulnerable.

Many states also screen newborns for cystic fibrosis, since there is good evidence that early diagnosis may lead to better medical care and nutrition.

The full list of conditions for which screening is recommended by the March of Dimes can be found at http://www.marchofdimes .com/professionals/14332_15455.asp. A particularly important entry in that list is a screen for hearing impairment, which affects about two or three of every 1,000 newborns. Congenital hearing impairment can be caused by numerous mutations or can have nongenetic causes. Without testing, babies with hearing impairment may not be diagnosed for many months, at which time speech and language development may have been injured in a way that is difficult to reverse.

The list of conditions for screening will get longer and longer as medical research advances. Newborn screening currently involves a few drops of blood from the baby's heel, absorbed onto a filter paper card, and then analyzed in a central laboratory. More recent technologies being employed in some states allow testing for a long list of disorders of amino acids, organic acids, and sugars, going even beyond the March of Dimes recommendations. At times this presents a real problem, in that a particular newborn may turn out to have a previously undescribed abnormality. Some of these conditions are harmless, but some could lead to mental retardation, or even be fatal. The challenge of handling these metabolic conditions of unknown significance can be quite vexing for the health care professional, and profoundly unnerving for parents. But despite this, there is no question that newborn screening provides a remarkable advance in the early identification of treatable genetic conditions.

Newborn screening seems almost certain to evolve into an even broader and more comprehensive survey. As the costs for sequencing the entire genome progressively fall, probably to less than $1,000 in the next five to seven years, arguments for obtaining that information at the time of birth will become increasingly compelling. For some of us, this may be a cause for anxiety. A scene in the 1997 film *GATTACA* depicts a high-tech delivery room where a genome analysis is done immediately after birth, complete with predictions of a precise and scary future outcome for the movie's hero. This is *not* our future: genes are generally not destiny, especially for common conditions like heart disease, diabetes, or cancer. But a softer version of *GATTACA* may be coming soon.

I will have more to say later in this book about genetic privacy and the right *not* to know about future risks. After all, once a DNA sequence has been determined, that individual has lost the opportunity to say, "No, thanks." On the other hand, as we learn more about effective interventions for genetic risk factors, and recognize that interventions early in life may provide significant advantages, it will become more and more compelling to determine this information at birth. One possible compromise is to create a method of obscuring unnecessary information until the individual reaches the age of 18 and can then decide what he or she wants to know.

Many people instinctively recoil at the vision of a future like *GATTACA*. But consider the case of obesity. This problem is highly heritable, with current estimates suggesting that roughly 60 to 70 percent of one's adult body weight is determined by genes. Several of these genes have already been discovered. If you gave birth to a baby with a high genetic risk of obesity, you could modify her diet from infancy, rather than waiting until she reached age five or later and was already overweight, and already in the habit of overeating.

Carrier Screening

With recessively inherited conditions, carriers are often entirely normal, but the child of two carriers has a one chance in four of being affected (Figure 2.2). The first major push to offer *carrier screening* was for Tay-Sachs disease, which appears primarily but not exclusively in individuals of eastern European Jewish (Ashkenazi) background. Infants with Tay-Sachs disease appear to develop normally for the first six months of life. But then, as a storage material that they cannot metabolize builds up in the brain, a relentless deterioration occurs, including blindness, deafness, and paralysis. Death generally occurs by age four or five. The disease is caused by the absence of an enzyme called hexosaminidase A. An enzyme test capable of detecting the approximately one in 30 Ashkenazi Jews who is a recessive carrier of this condition was developed in the 1970s.

After extensive community consultation, carrier screening was offered to the Jewish population in the 1970s. There was strong interest. Couples who were both found to be carriers for Tay-Sachs disease generally wanted to be informed, in order to make reproductive decisions to avoid the birth of a child with this terrible condition. Carrier couples were presented with several options: adoption; artificial insemination from a noncarrier donor; or, for those willing to consider it, prenatal testing combined with possible termination of the pregnancy.

The use of carrier screening in the Jewish population has been very high, and the number of Jewish children born with Tay-Sachs disease has plummeted to almost zero. Ironically, this disease now is seen primarily in children of other ethnicities, where the incidence of mutations is much lower but no screening program has been carried out.

Efforts were also made in the 1970s to institute carrier screening programs for sickle-cell anemia, since about one in ten African-Americans is a carrier. In this instance, the results were much less successful. Although the efforts were well intentioned and were supported by leaders in the African-American community, there was much confusion about the differences between having sickle-cell disease and being a carrier, often referred to as sickle-cell trait. Sickle-cell trait has essentially no consequences for the individual's health, except for extreme circumstances such as high-altitude plane flight in an unpressurized cabin. But that information was not always explained clearly. Even worse, although carrier couples could be readily identified by a simple screening test available in the 1970s, no effective prenatal test was available, so the options available to carrier couples were more limited than the options for Tay-Sachs disease. Those offering the screening were often white, and those being screened were usually black; this situation raised the specter of a eugenic agenda. Most carrier screening programs were ultimately shut down.

With the discovery of the genetic basis for cystic fibrosis, the opportunity to provide couples with information about their risk of having a child with CF has emerged. But this has not been without controversy. After all, survival of individuals with cystic fibrosis has been steadily improving (Figure 2.5), so this is a far cry from Tay-Sachs disease. Nonetheless, research studies in the 1990s indicated an interest on the part of couples in having the information.

Yet there is a major problem here. Our health care system is not set up to encourage screening of individuals or couples prior to conception. Currently, the first carrier test for CF is almost always performed at the first prenatal visit to the obstetrician, by which time a pregnancy is already under way. In the view of many of us, the strategy followed for Tay-Sachs disease, whereby carrier couples are identified prior to starting a pregnancy, would be much preferable, since it preserves more options for the couple. If I were younger and about to start a family, I would want to test myself and encourage my wife to do the same—not just for CF, but for a long list of recessive diseases. Currently, about one out of every 1,000 pregnancies in America involves a disease that could have been predicted by carrier screening. You would be surprised how often these diagnoses are a complete shock, since a recessive gene can be passed on for generations without any sign of its presence. But our current model of delaying carrier screening until a pregnancy is already under way forces couples to make tough choices, and deprives them of pre-conception alternatives that they might have preferred.

There is much more to say about carrier testing. How reliable are the tests? You may recall that whereas cystic fibrosis is virtually always caused by mutations in the *CFTR* gene, there are more than 1,000 different ways of misspelling this gene. In order to try to identify as many carriers as possible, without making the test inordinately expensive, most current carrier screens look for the 23 most common *CFTR* mutations, and in consequence they pick up about 90 percent of carriers. This means, however, that children with CF can still occasionally be born to a couple when one of the parents had tested negative.

Carrier screening should never be undertaken without the full informed consent of the individual for whom the test is being offered. This is particularly true when carrier screening is offered at the time of an existing pregnancy. After all, the identification of a pregnancy at risk for CF or some other recessive disease presents couples with difficult choices. If parents have no interest in getting the information about an affected child before birth—if, for example, they would never terminate a pregnancy under any circumstances—then it may be reasonable for them to decline this kind of testing.

It is important, however, to recognize that going through the testing does not necessarily imply a decision to terminate an affected pregnancy. For some couples, there may also be a desire to know the information in order to prepare for the birth of a child with special health care needs.

Short of having complete genome sequences for all couples prior to starting a pregnancy, are other carrier screening tests being contemplated in the near future? One that is under consideration is for spinal muscular atrophy (SMA). Infants affected with this recessive condition appear normal at birth but develop loss of muscle tone during their first few months of life, ultimately progressing to a complete flaccid paralysis and death by age two. About one in 40 individuals is a carrier for SMA; this means that one in 1,600 pregnancies is at risk, and the child will be affected in one-quarter of those. As with all recessive diseases, carriers have no symptoms and usually no family history of the disorder. Given the severity of the disease, carrier screening seems as desirable as with Tay-Sachs. Unfortunately, the genetic test is rather complicated. The basis of the disease is the deletion of an entire copy of a duplicated gene, and the presence of the mutation can be detected only by careful quantitative analysis of a relatively large snippet of DNA. The available test detects 94 percent of carriers, but it costs several hundred dollars.

Another carrier test on the table is for fragile X syndrome. This condition gets its name from the fact that the X chromosome of an affected male often has a visible fragile spot, which can be detected with a microscope after cells in a laboratory culture are treated with certain chemicals. Until 1990, this complicated and difficult analysis was the only test available. At that time, however, the molecular basis of the disease was uncovered. A particular gene on the X chromosome is inactivated in this condition. The cause is a tandem duplication of the sequence CGG just "upstream" of the gene. Normal individuals have fewer than 45 of these CGG repeats. If the repeat extends to more than 200 copies, it effectively shuts down the gene. Because this gene is located on the X chromosome, and females have two Xs whereas males have a single X, this disease affects boys far more than girls.

Figure 2.6 shows a typical example of X-linked inheritance. Females may be carriers of an X-linked recessive condition, but generally they are not affected because they also have a normal X. A male child of a carrier female has a 50 percent chance of being affected, but you should never see an instance of male-to-male transmission, since fathers must transmit their Y chromosome, not their X chromosome, to their sons.

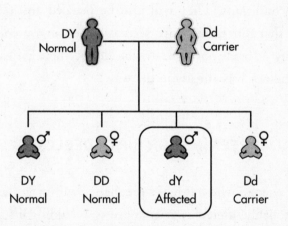

Figure 2.6: X-linked inheritance, where generally only males, with their single X chromosome, are affected. "D" is the normal copy of the gene, "d" is the abnormal copy.

Fragile X syndrome is the second most common cause of mental retardation, ranking second only to Down syndrome. Approximately one in 4,000 males is affected, and this condition occurs in all ethnic groups, often in families with no history of mental retardation. Furthermore, this condition is somewhat unusual for an X-linked recessive, in that about one-third of carrier females actually have mild learning disabilities or even mild mental retardation.

Given the significance of the condition, the frequency of carriers, and the availability of a DNA test (though it is not technically simple), there are increasing calls for offering fragile X carrier screening to all females. However, at the present time no consensus has been reached about proceeding down this pathway.

Debates about the appropriateness of carrier screening will be likely to change in character in the coming few years, as more and more individuals will have complete DNA sequences of their entire genome determined, revealing all of their carrier status risks and providing an opportunity for couples to know about those risks prior to initiating a pregnancy. It is likely that within a few decades people will look back on our current circumstance with a sense of disbelief that we screened for so few conditions. They will also be puzzled and dismayed, as I am now, that our health care system put so many couples in an unnecessarily difficult position, by not identifying their carrier status until a pregnancy was already under way.

MATERNAL SCREENING FOR BIRTH DEFECTS

If you are a woman who has had a pregnancy within the last 20 years, you were probably offered a variety of tests, including ultrasound and maternal blood tests, to assess the status of the fetus in the first or second trimester. Ultrasound scans are now routinely done several times in the course of a pregnancy, and can detect a wide range of anatomic abnormalities, including congenital heart defects. The blood tests are primarily designed to detect the presence of chromosomal disorders such as Down syndrome (due to an extra chromosome 21) or neural tube defects, which involve abnormalities of the spine ranging from milder cases of spina bifida to much more severe conditions such as anencephaly, in which the brain fails to form completely.

The screening tests available at the present time for neural tube defects and chromosomal disorders are indirect, assessing the levels of certain proteins in maternal blood, together with ultrasound examination of the fetus, to try to provide any warning signals of trouble. A screening blood test can be done between 11 and 13 weeks after the woman's last menstrual period; this test includes measurement of a specific form of the pregnancy hormone hCG and another protein, called pregnancy-associated plasma protein A (PAPP-A). The ultrasound component of this combined test looks for the thickness at the back of the baby's neck, since increased thickness is present more often when the fetus is affected with a chromosomal disorder.

Alternatively, a second-trimester screening test is offered to most women, and is done within 15 to 20 weeks after the last menstrual period. This test measures the levels of three or four substances in the mother's blood. The "triple screen" measures alpha-fetoprotein (AFP), hCG, and a pregnancy hormone called estriol. The quadruple (quad) screen also measures inhibin A, allowing detection of Down syndrome at about 80 percent sensitivity (that is, the test catches 80 percent of all cases). Both the triple screen and the quad screen can detect about 75 to 80 percent of spina bifida and close to 100 percent of anencephaly.

It is important to recognize that these are screening tests and do not provide a definitive diagnosis. These tests are plagued by false positives, which can create a great deal of parental anxiety. A common cause of a false positive result is that the fetus is either a few weeks older or younger than previously thought, but there are other reasons for a false positive test that remain mysterious. The long-term consequences of having induced this uncertainty about the health of the fetus are not to be dismissed lightly. No woman should undergo this kind of screening process without being fully informed of the purpose and possible outcomes. Unfortunately, screening is all too often presented in a cursory fashion, with the implication that any concerned parent would want to have this information. No doubt concerns about litigation also drive obstetricians to make sure that screening is done.

There will be major changes in maternal screening for birth defects in the near future. One of the most dramatic will be the advent of screening for fetal DNA in maternal blood, allowing chromosome abnormalities like Down syndrome to be detected much more directly. Attempts have been made over more than a decade to do this efficiently, but it is only in the last year or two that a practical and reliable method has emerged, which takes advantage of the fact that there is a small amount of free DNA and RNA derived from the fetus in maternal blood. These new methods can often directly detect the presence of Down syndrome at 12 to 14 weeks of gestation. This approach will still have to be validated in a large study, however, before it is offered more widely.

PRENATAL DIAGNOSIS BY CVS AND AMNIOCENTESIS

The maternal blood screening tests mentioned above are not definitive, and suspicion of a chromosome abnormality must currently be followed up by a test that obtains actual cells derived from the fetus to learn for certain whether the fetus is affected. However, whether to pursue diagnostic testing is a personal choice, and not all women choose to undergo further testing. The test most commonly performed for this purpose over the last 40 years is amniocentesis, generally carried out between 16 and 20 weeks of gestation. In this procedure, a needle is inserted through the abdomen into the fluid surrounding the fetus, and a moderate amount of that fluid is drawn off. Cells floating in the fluid that derive from the fetus can then be cultured in the laboratory and their chromosomes studied.

An alternative method of obtaining fetal cells, chorionic villus sampling (CVS), is also available. This involves inserting a catheter through the vagina, or a needle through the abdomen, to remove a small amount of the fetal portion of the placenta, and can be carried out at 10 to 12 weeks of gestation. Initially there were concerns that CVS might carry a higher risk of miscarriage, but with increasing experience it appears that the risk is no greater than one in 200. (Amniocentesis carries an estimated risk of one in 400.)

Traditionally, these methods for detecting chromosomal disorders were reserved for women over age 35, when chromosomal disorders are more common, or for women with abnormal screening results. More recently, however, the American College of Obstetrics and Gynecology has recommended that CVS or amniocentesis be offered to women regardless of age, depending on the wishes of the patient and the provider.

As methods of analyzing the genome become increasingly more sophisticated, it is possible to test cells derived by CVS or amniocentesis for more subtle mutations. Although these methods still fall short of complete genome sequencing, they are able to detect small deletions or rearrangements of chromosomes that are not visible in a standard microscopic analysis.

This capability can, however, be both a blessing and a curse, as such small chromosomal changes may be of uncertain significance. Once discovered, they can lead to an investigation of both parents to see whether the chromosome abnormality is inherited. If it is, and if the parent in question appears to be normal, then generally a sigh of relief is breathed, but only after a fair amount of parental anxiety has been incurred. When neither parent shows the alteration, so that it has clearly happened for the first time in the fetus, the consequences are often unpredictable, and very difficult decisions must be made.

PRE-IMPLANTATION GENETIC DIAGNOSIS (PGD)

The discovery of a fetus with a serious genetic anomaly in the second trimester of a pregnancy presents a wrenching dilemma. No parents who face that situation can avoid the powerful emotional and physical consequences. Faced with the desire to avoid that trauma, and building on the increasing success of in vitro fertilization (IVF) as a means of assisted conception, a dramatic new approach to prenatal diagnosis has been developed over the last decade.

Pre-implantation genetic diagnosis (PGD) depends upon the ability to bring sperm and egg together in the laboratory, following surgical harvesting of multiple eggs from the prospective mother after appropriate hormonal stimulation. After fertilization with the father's sperm, the resulting embryos are observed, and in three days will have reached the eight-cell stage. Remarkably, at this point an embryo "biopsy" can be carried out, removing one of the eight cells for diagnostic purposes, and yet still allowing normal development of the embryo from the remaining seven cells. Extremely sensitive DNA methods are necessary in order to obtain accurate DNA tests from this single cell.

After the DNA diagnosis is carried out on the available embryos, a decision is made about which ones to implant, assuring the couple that only those with the desired DNA test results will have an opportunity to go on to a full pregnancy. Note, however, that for those who believe that life begins at the moment of conception, this does not eliminate the moral concerns about termination of a pregnancy.

The initial motivation for PGD arose from the desire to prevent severe recessive diseases, such as Tay-Sachs disease. But over the past decade, the procedure has become more widely available, and has been applied to a wider and wider range of conditions. Those include such disorders as cystic fibrosis, and even testing for adult-onset conditions such as the presence of a *BRCA1* mutation that predisposes a woman to a high risk of breast and ovarian cancer. The potential extension of PGD to more and more non-crippling conditions has raised the specter of "designer babies."

In the United States there is no systematic method of collecting data or setting standards about the applications of PGD, but in the United Kingdom the Human Fertilization and Embryology Authority (HFEA) makes rules about the application of this method. In 2006, the HFEA agreed to consider the use of PGD for conditions such as inherited breast, ovarian, and bowel cancers when these are attributable to a single highly penetrant mutant gene (such as *BRCA1*). The HFEA stated, "The decision . . . deals only with serious genetic conditions that we have a single gene test for. We would not consider mild conditions like asthma and eczema, which can be well managed in medical practice. We would not consider conditions like schizophrenia, where a number of genes have been identified, but there is no single gene that dictates the condition."

Lacking regulation in the United States, however, PGD is finding increasing applications to circumstances that are less and less consistent with such principles. In fact, a recent survey indicates that 42 percent of clinics offering PGD in the United States would be willing to apply this procedure solely for the purpose of sex selection. One California lab recently advertised the availability of PGD for selection of eye and hair color! The resulting public outcry caused rapid withdrawal of the offer, and the scientific basis of such predictive abilities was shaky at best, but a shiver went through the more socially responsible segments of the IVF community. All this conjures up another scene from the movie *GATTACA*, in which widespread utilization of PGD is offered, in fact mandated, in order to "maximize the potential" of the offspring of all couples. In one memorable scene from that movie, a smooth-tongued medical geneticist shows a prospective couple a series of their potential embryos that have been derived by

in vitro fertilization and PGD, and argues that there is nothing unethical about the procedure. "It's still you," he says, "simply the best
of you. You could conceive naturally a thousand times and never get
such a result."

The scene is chilling. Is that where we are headed? But the premise
is scientifically flawed. The couple are told that this embryo selection
will optimize a wide variety of characteristics of their future offspring,
including intelligence, athletic ability, musical talent, and physical at-
tractiveness. Yet we know that all these conditions are affected by a
long list of individually weak genetic contributors, and the environment is hugely important to the ultimate outcome. Imagine, for instance, that there are 10 different genes that influence each of four
characteristics that the parents wish to optimize. To optimize for all
these outcomes, even assuming that at least one of the parents carried
the desirable variations, would require *billions* of embryos. I do not
dismiss the moral concerns of PGD—indeed, I believe this procedure
is already misapplied for sex selection in the United States. I merely
offer a scientific limit to the plausibility of the *GATTACA* scenario.
A couple hoping for a son who will play first violin in the orchestra,
earn an A+ in math, and be the quarterback on the football team
might discover instead that they are the parents of a sullen 15-year-old
who stays in his room listening to heavy metal music, smoking pot,
and cruising the Internet for pornography or the latest violent video
games. In other words, DNA testing will never be a match for commitment to parenting.

A particularly controversial application of PGD has arisen for
couples who already have a child with a serious disorder, for which
a bone marrow transplant is desperately needed. Lisa and Jack Nash
found themselves in this situation, when their first child, Molly, was
born with a condition called Fanconi anemia, a recessive disease that
results in failure of the bone marrow to produce red cells and white

cells. While searching for a possible bone marrow donor for Molly, Lisa and Jack decided to attempt a second pregnancy, and heard about PGD as a method of ensuring that the next child would not be affected with Fanconi anemia.

In a discussion with genetic experts, however, the possibility came up of selecting an embryo that was not only free of Fanconi anemia but also a transplantation match for Molly. That opportunity was particularly appealing, since stem cells derived from the umbilical cord of a newborn can be obtained quite easily and without risk to the infant. An extensive ethical debate ensued. Was it appropriate for Lisa and Jack to plan the birth of a child who would not only be valued for his own sake, but also serve as a much-needed tissue donor for his sister?

Ultimately, the decision was made to proceed, and Mrs. Nash underwent four cycles of in vitro fertilization, each time choosing embryos to achieve both goals: a disease-free child and a donor for Molly. Ultimately, a pregnancy was begun, and nine months later Adam Nash was born. His stem cells served as the transplant source for his six-year-old sister. When last heard from, both children were doing well.

CONCLUSION

In this chapter we have considered conditions in which genetics plays a particularly powerful role. In those circumstances, inheritance follows rather predictable statistical rules, DNA tests can be fairly definitive, and outcomes are reasonably predictable. Collectively, such genetic disorders account for 5 to 10 percent of pediatric hospital admissions and touch the lives of many individuals and families, but individually these are relatively uncommon conditions. Finding such a condition in yourself or your family—or even the potential for its occurrence—can have powerful consequences, and the diagnosis and treatment of these diseases will remain a major component of personalized medicine.

Until recently, the story of genetic medicine would end here. This book would be primarily aimed at the small fraction of us who face those diseases. Yet the revolution in human genetics is extending rapidly beyond these less common conditions to reveal the role of individual genetic factors in much more common conditions such as diabetes, heart disease, and cancer.

To put this another way, until now we have dealt with errors in the language of life that even a relatively inexperienced reader could detect. Now we are going to move on to a new world of more subtle linguistic mysteries.

WHAT YOU CAN DO NOW TO JOIN THE PERSONALIZED MEDICINE REVOLUTION

If you are a considering becoming a parent at some time in the future, talk to your doctor about screening both yourself and your partner prior to conception. If either of you has any family history of a serious inherited condition, or if you come from a population where a disease such as cystic fibrosis (northern Europeans), Tay-Sachs disease (Ashkenazi Jews), thalassemia (Mediterraneans or Southeast Asians), or sickle-cell anemia (West Africans) is common, this might be information you would benefit from knowing before starting a pregnancy.

CHAPTER THREE

Is It Time to Learn
Your Own Secrets?

Sergey Brin thought DNA testing was mostly entertainment. As a cofounder of the enormously successful search engine Google, he had helped transform the world's access to information, making the Internet a clearinghouse and a meeting place for vast numbers of people in ways not previously imagined. So when his wife, Anne Wojcicki, invited him to be one of the first individuals to undergo genome-wide testing as part of her startup company, 23andMe, he agreed. He even encouraged other family members to participate, and was able to enjoy seeing which parts of his DNA he shared with various relatives. The risk predictions for disease showed that he was at slightly lower risk for some, perhaps a little higher for others, but he concluded that there were no big surprises.

This all changed when 23andMe offered a new service, and his wife encouraged Sergey to look at one particular genetic variant in a gene called *LRRK2*. You see, Sergey's mother had been diagnosed with Parkinson's disease, and recent research had shown that in some instances an uncommon mutation in the *LRRK2* gene can confer a high risk of this late-onset neurological condition. Looking closely at the results from the DNA chip that 23andMe uses for its genetic testing service, Sergey discovered that both he and his mother carry that *LRRK2* mutation. Genetic testing wasn't just entertainment after all.

This was a dramatic and unexpected result, as most of the genetic risk factors tested by this direct-to-consumer company are common and convey rather modest risks. But this mutation predicts Sergey's risk of Parkinson's disease to be roughly 74 percent by age 80. Sergey is still young, and by the time he is old enough for such symptoms to become a real problem, there might be a means of treatment—but that is certainly not guaranteed. Sergey wrote the following in his blog: "This leaves me in a rather unique position. I know early in my life something I am substantially predisposed to. I now have the opportunity to adjust my life to reduce those odds. I also have the opportunity to perform and support research into this disease long before it may affect me. And, regardless of my own health, it can help my family members as well as others. I feel fortunate to be in this position. Until the fountain of youth is discovered, all of us will have some conditions in our old age, only we don't know what they will be. I have a better guess than almost anyone else for what ills may be mine—and I have decades to prepare for it."

Welcome to the genome era. Now, what do YOU want to know?

FINDING THE TICKING TIME BOMBS IN EACH OF OUR GENOMES

Sergey's story is stunning, but not unique. Each of us is at risk for dozens of conditions, which we will either experience or not—as a result of the combination of risk factors we have inherited, and whether or not we encounter the environmental trigger that sets the disease process in motion. *There are virtually no conditions for which heredity does not play a role.*

Perhaps you will protest about this sweeping statement. After all, being hit by a brick dropped from the roof of a building was probably not influenced by your heredity—but it might reflect the heredity of the person who dropped the brick. And your genes will most certainly influence your ability to recover from the injury.

Your own risk factors for disease can be partially deduced by a careful medical family history, and you should most certainly take advantage of that "free genetic test" to assess your own risks. But not all individuals know their family history (that's especially true for those who are adopted), and even the most complete family history in an era of relatively small families may not reveal all of the risk factors, since the inheritance patterns for common diseases are often complicated and unpredictable.

We have already discussed those relatively rare conditions that are highly heritable, in which mutations in a single gene can predictably cause illness. But the genomic revolution is now extending to more common diseases and conditions. Diabetes, the common cancers, heart disease, stroke, and mental illness do not follow simple inheritance patterns, but they are still heavily influenced by genetics.

Understanding those disorders requires a more complex model of inheritance. Recent discoveries place us in a position to make several strong statements: (1) for each disease, specific genetic and environmental risk factors exist, and are rapidly being identified; (2) these discoveries are providing powerful new insights into both treatment and prevention; (3) the more you know about all this, the more you can adjust your own lifestyle and medical surveillance to prevent illnesses or catch them in early and treatable stages.

By way of illustration, let us focus on adult-onset (type 2) diabetes. Clearly, there must be genes involved, as the sibling of someone affected with type 2 diabetes (T2D) has his or her baseline risk increased by a factor of three. Looking at families in which T2D has afflicted more than one individual, however, makes it clear that a dominant, recessive, or

X-linked pattern cannot explain the occurrence of the disease—there is no single gene for T2D. Rather, there must be dozens of genetic variants that predispose to this condition, with each variant contributing only a small risk. Geneticists call this kind of inheritance *polygenic*. Each of us has a collection of such variants that may elevate our risk of T2D above the average, keep us in the middle zone, or reduce our chances of coming down with this disease. Someone with a high genetic load may develop diabetes even with relatively little in the way of environmental triggers. Someone with a moderate risk may develop T2D only if other factors such as weight gain, lack of exercise, and diet push him or her over the threshold into actual illness. Someone with a very low genetic risk may be able to avoid T2D, even with lifestyle choices that are unhealthy. So it is a combination of the genes that you have inherited and the environment that you live in that determines the outcome. Hence the common saying, "Genes load the gun, and environment pulls the trigger."

The multiplicity of genetic risk factors for diseases like diabetes, cancer, or heart disease has made this an extremely difficult detective story for researchers, who are attempting to identify the culprits in DNA. The strategies that worked so well for single-gene disorders like cystic fibrosis were found to be woefully underpowered for these polygenic conditions. Frustrated by the failure of family-based analysis, researchers tried other shortcuts. One of those approaches involved the so-called candidate gene strategy: try to guess which of the 20,000 human genes is involved in a particular disorder, and look for variations in that gene in affected people.

Perhaps you've heard the joke about the guy whose keys fell out of his coat pocket late at night on a darkened street. Realizing later that he was without his keys, he began searching. His companions were surprised to find him searching vainly in just one place—under the streetlight. Asked why he was limiting his search, he explained, "Anyone knows you can't find your keys where there's no light." Sadly, the candidate gene strategy generally suffered the same fate—and we didn't find the keys.

Frustration mounted. Very few genetic risk factors for common diseases like T2D had been identified by 2003. But the genome is a bounded set of information. Why couldn't this be done more systematically? Why couldn't we light up the whole street?

So let's do a thought experiment. Suppose you have DNA from 1,000 individuals who have diabetes and 1,000 otherwise well matched individuals who clearly do not. And now suppose you determine the complete DNA sequence for each of those 2,000 people, and then you compare them side by side (Figure 3.1). No longer are you limited to looking at candidate genes; you can look at the whole genome. You would still have to separate signal from noise. Some people might develop diabetes because of two or three rare mutations, others because of a larger set of common mutations. But assuming you have done the sequencing accurately, you should discover the major risk factors. You should even be able to say something about how important each variant is, basing your assessment on how often you see it in cases (the people with diabetes) and how often you see it in controls (the people who don't have diabetes).

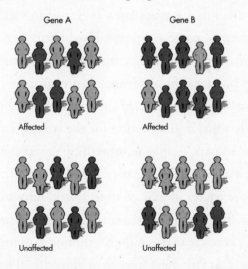

Figure 3.1: Finding the genetic glitches associated with risk of disease requires identification of variations in DNA that appear more commonly in affected than unaffected persons. Here gene B looks as if it might harbor an important risk factor, whereas gene A doesn't seem to be involved.

In 2003, when the Human Genome Project was completed, it seemed that a long time would be needed before this thought experiment could actually be performed. At the time, we did not even know the location of the 10 million common variations in the genome, much less the rare ones. But focusing initially on the common ones made sense. Most of these are single base-pair differences between individuals. These are called "single nucleotide polymorphisms" (SNPs). In the second place, the cost of a laboratory test to assess the actual DNA sequence at one of these variants in one DNA sample (a process called "determining the genotype") was about 50 cents in 2003. Thus the cost of the laboratory work to scan the entire genome for common SNPs associated with disease—called a genome-wide association study (GWAS)—on 1,000 cases and 1,000 controls would be 10 million SNPs times 2,000 DNA samples times 50 cents per SNP, or $10 billion! This was obviously a complete nonstarter. (And that's not even a full sequencing, just an analysis of the common variations.)

As I write this just six years later, it is nothing short of astounding that what seemed to be an impossibly difficult task has now been carried out for dozens of diseases, involving more than 100,000 DNA samples. Rarely has a technical advance in any aspect of science moved so quickly. That advance has fueled the excitement about personalized medicine, and is a major justification for the writing of this book.

It happened, in part, thanks to something called the *HapMap* project. Figure 3.2 displays approximately 2,000 letters of the DNA code. For most of those letters, nearly all of us would have exactly the same DNA sequence. But in three locations, common variations occur in this sequence. These are three of the 10 million common SNPs in the human genome. But you don't have to analyze all 10 million of them, one by one, to seek the causes of, say, diabetes. It turns out that SNPs are social animals, and they travel in packs.

```
GAAATAATTAATGTTTTCCTTCCTTCTCCTATTTTGTCCTTTACTTCAATTTATTTATTT
ATTATTAATATTATTATTTTTTGAGACGGAGTTTCACTCTTGTTGCCAACCTGGAGTGCA      1
GTGGCGTGATCTCAGCTCACTGCACACTCCGCTTTC C/T GGTTTCAAGCGATTCTCCTGC
CTCAGCCTCCTGAGTAGCTGGGACTACAGTCACACACCACCACGCCCGGCTAATTTTTGT
ATTTTTAGTAGAGTTGGGGTTTCACCATGTTGGCCAGACTGGTCTCGAACTCCTGACCTT
GTGATCCGCCAGCCTCTGCCTCCCAAAGAGCTGGGATTACAGGCGTGAGCCACCGCGCTC
GGCCCTTTGCATCAATTTCTACAGCTTGTTTTCTTTGCCTGGACTTTACAAGTCTTACCT
TGTTCTGCCTTCAGATATTTGTGTGGTCTCATTCTGGTGTGCCAGTAGCTAAAAATCCAT
GATTTGCTCTCATCCCACTCCTGTTGTTCATCTCCTCTTATCTGGGGTCAC A/C TATCTC      2
TTCGTGATTGCATTCTGATCCCCAGTACTTAGCATGTGCGTAACAACTCTGCCTCTGCTT
TCCCAGGCTGTTGATGGGGTGCTGTTCATGCCTCAGAAAAATGCATTGTAAGTTAAATTA
TTAAAGATTTTAAATATAGGAAAAAAGTAAGCAAACATAAGGAACAAAAAGGAAAGAACA
TGTATTCTAATCCATTATTTATTATACAATTAAGAAATTTGGAAACTTTAGATTACACTG
CTTTTAGAGATGGAGATGTAGTAAGTCTTTTACTCTTTACAAAATACATGTGTTAGCAAT
TTTGGGAAGAATAGTAACTCACCCGAACAGTGTAATGTGAATATGTCACTTACTAGAGGA
AAGAAGGCACTTGAAAAACATCTCTAAACCGTATAAAAACAATTACATCATAATGATGAA
AACCCAAGGAATTTTTTTAGAAAACATTACCAGGGCTAATAACAAAGTAGAGCCACATGT
CATTTATCTTCCCTTTGTGTCTGTGTGAGAATTCTAGAGTTATATTTGTACATAGCATGG
AAAAATGAGAGGCTAGTTTATCAACTAGTTCATTTTTAAAAGTCTAACACATCCTAGGTA
TAGGTGAACTGTCCTCCTGCCAATGTATTGTCACATTTGTGCCCAGATCCAGCATAGGGTA
TGTTTGCCATTTACAAACGTTTATGTCTTAAGAGAGGAAAATATGAAGAGCAAAACAGTGC
ATGCTGGAGAGAGAAAGCTGATACAAATATAAATGAAACAATAATTGGAAAAATTGAGAA
ACTACTCATTTTCTAAATTACTCATGTGATTTTCCTAGAATTTAAGTCTTTTAATTTTTGA
TAAATCCCAATGTGAGACAAGATAAGTATTAGTGATGGTATGAGTAATTAATATCTGTTA
TATAATATTCATTTTCATAGTGGAAGAAATAAAATAAAGGTTGTGATGATTGTTGATTAT
TTTTTCTAGAGGGGTTGTCAGGGAAAGAAATTGCTTTTTTTCATTCTCTCTTTCCACTAA
GAAAGTTCAACTATTAATTTAGGCACATACAATAATTACTCCATTCTAAAATGCCAAAAA
GGTAATTTAAGAGACTTAAAACTGAAAAGTTTAAGATAGTCACACTGAACTATATTAAAA
AATCCACAGGTGGTTGGAACTAGGCCTTATATTAAAGAGGCTAAAAATTGCAATAAGAC
CACAGGCTTTAAATATGGCTTTAAACTGTGAAAGGTGAAACTAGAATGAATAAAATCCTA      3
TAAATTTAAATCAAAAGAAAGAAACAAACT A/G AAATTAAAGTTAATATACAAGAATATG
GTGGCCTGGATCTAGTGAACATATAGTAAAGATAAAACAGAATATTTCTGAAAAATCCTG
GAAAATCTTTTGGGCTAACCTGAAAACAGTATATTTGAAACTATTTTTAAACCGAGTTAT
GGCACACTTGGGCAATTTCAGAGATT
```

The boxes mark three common variants (SNPs).

But it turns out they are closely correlated:
C at SNP 1 is always found with A at SNP 2 and G at SNP 3.

Figure 3.2: Two thousand letters of the DNA code (showing just one strand).

Apparently, the relatively recent origin of our species from a relatively small group of founders has resulted in a limited number of
chromosomal types, which we refer to as *haplotypes*. Our genetic
variation is therefore not properly thought of as an independent

collection of 10 million common differences. Instead, these differences are organized into local neighborhoods. Knowing one or two variations in each neighborhood allows you to predict what would have happened if you had tested the others. Some of the neighborhoods are quite small, and others extend across large stretches of DNA. On the average, 30 or 40 SNPs travel together in one of these neighborhoods.

If you know the boundaries of the neighborhoods, and you choose a surrogate set of SNPs wisely, you can be comprehensive without spending a huge sum of money on the laboratory work to test every one of them.

The HapMap project was designed to define the boundaries of the neighborhoods, providing a shortcut to genome analysis that reduces the amount of work by about a factor of 40. I had the privilege of serving as the project manager for the HapMap, a highly organized and rapidly moving international program that brought together more than 2,000 scientists in six countries to produce the catalog of human variations and organize it into neighborhoods, making all the data immediately available to the public for free.

The other dramatic development since 2003 has been the profound drop in costs of genotyping, from 50 cents to a fraction of a penny. This was inspired by a variety of highly creative approaches, many of them marrying the technology of computer chips with DNA chemistry to produce "DNA chips" that were capable of assessing as many as 1 million SNPs in one array the size of a postage stamp.

With HapMap reducing the number of SNPs that needed to be tested, and with the cost of genotyping dropping profoundly, a genome-wide association study of 1,000 cases and 1,000 controls could be conducted by 2006 for substantially less than $1 million. What a dramatic change in just a few years!

THE FIRST SUCCESS STORY: MACULAR DEGENERATION

My aunt Martha was a bit of a character. Brilliant, opinionated, and widely read, she ultimately rose to become headmistress of a private school where the students both admired and feared her. As her nephew and her godson, I shared both of those reactions. She was a great teacher, dedicated to her students, but her strong personality and Julia Child voice were enough to intimidate a small boy. Sadly, one of her greatest enjoyments in life was significantly compromised during her retirement, as she developed progressive visual difficulties, eventually diagnosed as age-related macular degeneration. Starting in her late seventies, and progressing to almost complete blindness in her last years, this cruel disease robbed her of the joy of reading.

Most of us did not expect the cause of macular degeneration to be found lurking in DNA. After all, a disease that arises at age 70, 80, or 90 would hardly seem likely to be much influenced by heredity— we generally tend to associate stronger hereditary factors with earlier-onset conditions. But in 2005, using early data from the HapMap project, researchers at Yale University, studying just 96 affected individuals, were able to determine that a totally unexpected gene harbors a common variation that plays a major role in risk for this disease.

Shortly after that, another unexpected gene on a different chromosome with a similarly large effect on macular degeneration was identified. It became clear that almost 80 percent of the risk could be inferred from a combination of these two genetic risk factors, combined with just two environmental risk factors (smoking and obesity).

These results electrified the scientific community. Until this success story, there had been considerable skepticism about whether the HapMap strategy was going to work. These findings for macular degeneration, immediately confirmed by multiple groups, laid that skepticism to rest.

This discovery has also prompted a whole new approach to treatment. Both genes identified for macular degeneration are involved in the inflammatory pathway, suggesting that inflammation may be playing a much more important role than was previously understood. Many drugs have already been developed that target the inflammatory system, so the possibility of applying those for prevention or treatment of macular degeneration is now a very hot topic. Interestingly, it had been previously noted that individuals with rheumatoid arthritis, who receive high doses of anti-inflammatory agents to try to control their arthritis, have a very low incidence of macular degeneration. Could that have been a clue all along? Clinical trials will be needed to answer these questions, but such trials would never have moved in this direction had it not been for the stunning discoveries coming out of the human genome.

These discoveries about macular degeneration have taken on particular personal significance for me, since the survey of my own DNA has revealed a higher risk for this condition. Am I headed for the same outcome as Aunt Martha? I'm not ready to start popping a dozen Advil tablets a day just yet, as that might do damage to my kidneys or my stomach lining—so I'll wait to see what the clinical trials say about the value of that approach. But as I mentioned in the introduction, the suggestion that dietary omega-3 fatty acids might provide some protection against macular degeneration has caused me to make sure to include these items, especially oily fish, in my own diet. While the evidence that this will help is somewhat sketchy, there is certainly no evidence that it could be harmful—and I rather like fish anyway.

THE DELUGE OF DISCOVERY

What began as a trickle of insights about genetic risk factors in common diseases turned into a flood by 2007, and as of this writing the flood continues (Figure 3.3). Genetic risk factors for diabetes, heart disease, the common cancers, asthma, stroke, obesity, high blood pressure, and even atrial fibrillation and gallstones appeared in dizzying numbers, filling up the pages of the most prestigious biomedical research journals. Month after month, new and startling revelations were published and almost immediately confirmed by other groups. Nearly all the genes found to be involved in common diseases were surprises.

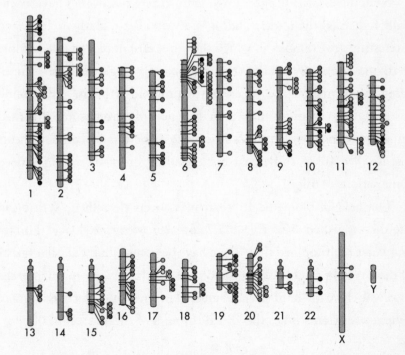

Figure 3.3: New findings about genetic risk factors for disease. Each symbol represents a newly discovered variant that predisposes to one of several dozen human conditions that are common in the population. This diagram would have had only seven entries in 2002.

An additional surprise emerged: for most of the genetic variants that played a role in risk of disease, the problem wasn't that the glitch led to a garbled protein; rather, the glitch affected whether the responsible gene was turned "off" or "on" at the right time and in the right amount.

Another common theme that began to emerge about genetic risk factors and disease is that each common variant individually contributes quite modestly to risk. Macular degeneration, in which just two genes play a very substantial role in the future risk of disease, is so far just about the only example where such large effects are contributed by common variants. For the most part, the variants being discovered increased the risk by only 10 to 40 percent.

Nonetheless, the insights provided by these discoveries have essentially redefined the understanding of illness. Particularly exciting and interesting are examples in which the same variant plays a role in more than one disease. For example, variations in a single gene were implicated in juvenile (type 1) diabetes, rheumatoid arthritis, and Crohn's disease. And variations in one small region of chromosome 9 turned out to play independent roles in type 2 diabetes and coronary heart disease. We are learning that our classification of disease may need some serious revision!

The field of biomedical research has been electrified. Writing in the distinguished *New England Journal of Medicine*, David Hunter and Peter Kraft opined that "there have been few, if any, similar bursts of discovery in the history of medical research." *Science* magazine, the most widely read "hard science" journal in the world, declared human genetic variation studies the "Breakthrough of the Year" for 2007.

WHAT DOES IT MEAN TO DISCOVER GENETIC RISK FACTORS? BACK TO DIABETES

Type 2 diabetes (T2D) currently affects 16 million people living in the United States, and approximately 150 million worldwide. It comes on insidiously. In fact, many individuals are not even aware they have the disease for several years after its onset. However, if left untreated, T2D has the potential for serious complications, including heart attack, stroke, blindness, kidney failure, and peripheral arterial disease that can require amputations.

The important organ systems involved in diabetes include the islet cells of the pancreas (which make insulin), the liver, muscle, the brain, and fat. Together these systems smoothly regulate insulin and glucose levels during periods of feeding and fasting. Diabetes results from a derangement of this balance, so that the amount of insulin being produced is insufficient for the glucose in the system.

Type 1 diabetes (T1D) is actually a very different disease. In this condition, the immune system attacks the beta cells of the pancreas, which normally secrete insulin, and ultimately destroys them. T1D commonly appears in childhood but can appear in older individuals. It is generally not associated with obesity. To be effectively treated, T1D requires insulin injections, since it is lack of insulin that presents the primary difficulty.

In type 2 diabetes, there is no immune attack on the beta cells. Instead, obesity is a major contributor. As the body gains weight, the demands for insulin increase. The beta cells of the pancreas, asked to produce substantially more insulin, ultimately become exhausted and begin to fall behind. The glucose levels in the blood climb. In a vicious circle, elevated glucose is also toxic to the beta cells. Ultimately, diabetes occurs. Successful treatment can be achieved in many cases with oral medications that stimulate the release of insulin from the beta cells that are still functioning, but in more advanced cases insulin has to be given by injection, just as in type 1 diabetes.

Thanks to the ability to search the entire genome, more than a dozen genetic risk factors have now been identified for type 1 diabetes. Although some of those genes are involved in the immune response, as might be expected, a number of others suggest wholly new approaches to prevention and treatment of T1D.

For type 2 diabetes, more than 20 genes have been identified as playing a role, and the number grows almost by the week. We still don't know what all those genes do, but for the half we do understand, nearly all point to the beta cell as the primary problem.

When the target genes for T2D emerged, it was striking to note that two of them code for the known targets of two of the most commonly used types of diabetes drugs. Of course these drugs had been arrived at by a completely independent approach, but the fact that these two genes appeared in a survey of hereditary factors almost certainly suggests that there are other targets on that list for drugs yet to be developed. These might actually turn out to be highly beneficial in the prevention and treatment of this common and devastating disease.

RISK PREDICTION AND THE RBI RULE

Developing new treatments for diabetes has to be considered a good thing. But for this disease and many others, the question newly arising from personalized medicine is: do you really want to know about your possible future? It's time for us to take a serious look at this question. In some instances, knowing your risks could save your life. But are there certain kinds of questions that you would rather leave unanswered?

It is useful at this point to reflect upon some general principles about risk prediction and human health. There are three major factors that, consciously or not, people generally use to assess whether or not they want this kind of information:

Factor 1. How big is the risk? To answer this, it is critical to keep in mind two different kinds of risks. You will often see cited a "relative risk," which describes whether your risk is higher or lower than that of the average person; a relative risk of 1.0 means that you're average, a relative risk of 0.5 means your risk is half that of the average person, and a relative risk of 1.5 means your risk is 50 percent higher than average. But most people will also want to know the "absolute risk" over a lifetime, in order to make an assessment about whether these predictors are meaningful. Both are important. For instance, if my relative risk of multiple sclerosis is 10 times normal, that sounds scary. But if the baseline risk of disease for the average person is only 0.3 percent (3 out of 1,000), then my risk is just 3 percent—meaning that there is a 97 percent chance that I will *not* develop multiple sclerosis. Thus the ominous-sounding relative risk of 10 is not all that significant for me personally.

Factor 2. What is the burden of the disease? Generally people are much more concerned about serious and potentially life-threatening diseases than about nuisance conditions. If you tell me my risk of cancer is substantial, you've got my attention. If it's tennis elbow, I might be concerned (especially if I'm Roger Federer), but not to the same degree.

Factor 3. What can I do about it? This is a critical component of the assessment that each of us must make as we decide whether or not we want to learn about particular risk factors for future disease. If you're going to tell me about my risk of heart attack, for which I can take preventive steps, I'm going to be much more interested than if you tell me about a risk for Alzheimer's disease and there is nothing definite that I can do about it. (Though even with Alzheimer's, it might change the way I want to plan my retirement.)

In effect, what each of us will do mentally when sizing up whether or not we want to receive this kind of risk information is carry out a simple mental multiplication. It looks like this:

Desire to Know = Risk × Burden × Intervention

Let's call this the RBI rule. It's a good one to keep in mind for your own health care and that of your family, whenever you are presented with an opportunity to learn information about risks in order to practice better prevention.

Now let's come back to diabetes. Of the more than 20 identified risk factors for T2D, the one with the strongest relative risk is a variant (an allele) in a gene called *TCF7L2*. This carries a relative risk of 1.4. But what does that mean? Again, you want to know the baseline risk. Distressingly, in the United States, that now stands at 23 percent by the age of 60. In other words, about one in four individuals will be diagnosed with diabetes by age 60. This figure will probably rise in the future, unless something is done about the current epidemic of obesity. Now, if you have the risk allele of *TCF7L2*, your risk of T2D goes up by a factor of 1.4, bringing it to 32 percent (23 × 1.4), or about one out of three.

The following table shows the relative and absolute risks for T2D, just based on this one gene:

	Average	With Risk Allele
Relative Risk	1.0	1.4
Absolute Risk	23%	32%

For most common diseases, including T2D, there are a multiplicity of such risk factors. To explain such relative risks, these are often depicted in a graphic form, as shown in Figure 3.4. Here each genetic risk factor is shown as having a relative risk above or below the average.

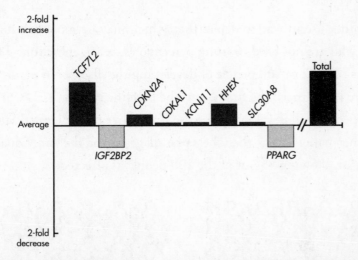

Figure 3.4: A typical graphic display of the results of genetic testing of an individual test for risk of type 2 diabetes. Eight different gene variants have been tested; each provides a risk that is higher (*TCF7L2, CDKN2A, HHEX*), lower (*IGF2BP2, PPARG*), or the same (*CDKAL1, KCNJ11, SLC30A8*) as an average person. The total risk for this person is elevated by a factor of 1.5.

A major question scientifically is whether these relative risks interact with each other in some complicated way, or whether the total risk to the individual can simply be derived by multiplying the individual genetic risk factors together. At least in some instances in other species, genetic risk factors seem to interact in a highly synergistic fashion, so that, for example, if you have risk alleles in A and B, your risk is substantially more significant than just the multiplication of their risk factors. Thus far, however, there is no strong example of a common human disease for which this is true. So multiplication is generally applied without really knowing if the risks could be greater (or, possibly, lower). In Figure 3.4, the total risk for all relevant genetic risk factors for a hypothetical individual has been arrived at by multiplication, and it comes out to 1.5.

But this is just the relative risk. With a relative risk of 1.5, this person faces an absolute risk of 35 percent (the baseline of 23 percent times 1.5).

Another useful way to depict this, which may be more helpful for people who are not used to using percentages, is a graphic image that portrays the risk to 100 people of developing the disease. In Figure 3.5 there are two panels: the first shows the baseline risk for T2D: 23 out of 100 people will develop the disease if they live in the United States. The other panel shows that the same diagram for the individual in Figure 3.4, showing 35 out of the 100 people as affected.

Average risk:
23 out of 100

Risk to individual:
35 out of 100

Figure 3.5: Another way of showing the 1.5-fold elevated diabetes risk for the person whose genetic test results were shown in Figure 3.4.

Returning to the RBI equation for our hypothetical individual, we have deduced that for type 2 diabetes the information that can be determined by DNA testing for risk factors (R) is modest, 1.5, and the absolute risk is 35 percent. But what about the burden (B)? There is no question that diabetes is an illness with potentially severe and life-threatening consequences, though the fact that these occur slowly over many years has perhaps caused diabetes to receive less attention than heart attack or cancer. Yet diabetes is listed as the seventh leading cause of death in the United States, and this probably underestimates its contribution to early death, since diabetes is listed on the death certificate of only about 35 to 40 percent of individuals who have diabetes at the time of death. Overall, the risk of death at any particular age among people with diabetes is about twice that of people without this disease. This disease should therefore be taken with great seriousness, and so the B factor in our RBI equation is quite high.

What about I, the possibility of interventions? First let's make one thing clear. Preventing a serious disease will generally not be associated with a simple, one-time-only, inexpensive, and easily accomplished intervention. Tracy Beck, who has the disease PKU and whom we met in Chapter 2, takes care of her prevention strategy every time she eats a meal. The families we heard about with long QT syndrome practice prevention every day by taking medication or even by undergoing a surgical procedure to implant a defibrillator. Nonetheless, those of us facing the potential of a severe illness ought to have the chance to practice prevention, even if that requires considerable personal investment.

Diabetes ranks high up on the list of preventable diseases. We know that obesity, diet, and exercise play a profoundly significant role. One need only look at the statistics about the rapid increase in the incidence of type 2 diabetes over the last few decades to appreciate that this increase must be based on environmental influences, since the gene pool could not have altered this much in such a short period of time.

The evidence that diabetes can actually be prevented in susceptible individuals was provided most dramatically by a large prospective study called the Diabetes Prevention Program (DPP). Individuals who were overweight and who had levels of blood glucose higher than normal (but not sufficiently high to warrant a diagnosis of diabetes) were randomly assigned to one of three interventions. One group, the lifestyle intervention group, received intensive training in diet, physical activity, and behavior modification. By exercising 30 minutes a day, five days a week, and limiting their intake of fat and calories, the participants in this group aimed to lose 7 percent of their body weight and to maintain that loss. Another group received a beta-cell stimulating drug, metformin. A third group received a placebo pill. The second and third groups were given information about diet and exercise, but no intensive intervention or motivational counseling was provided.

The results were so dramatic that the study was stopped early. Participants in the lifestyle intervention group reduced their risk of developing full-blown diabetes by 58 percent. That was true across all ethnic groups, and for both men and women. The results were most significant for participants aged 60 and older, who reduced their risk by 71 percent. In contrast, those receiving metformin reduced their risk by just 31 percent.

These dramatic results indicate that the "I factor" for type 2 diabetes is significant, and that this should be largely a preventable disease. Of course, a skeptic may argue that the interventions practiced by the successful group are simple commonsense actions that we all should take. Perhaps diabetes is just the cost of our couch-potato lives. Still, advocates for this kind of testing argue that information about an elevated genetic risk may cause people to take actions they otherwise would have ignored.

The suggestion that this kind of genetic information could be widely used for prevention has been called premature by some. But is this really so different from the familiar use of cholesterol to gauge the risk of heart disease? After all, your blood cholesterol is a function both of your genes and of your diet, with genes playing a strong role in your "set point." Many of us have become quite used to the idea of following cholesterol levels as a predictor of future heart attack risk. Interventions—dietary modification and drug therapy—are available to bring down a high cholesterol level. In many instances, dietary modification turns out to be insufficient to achieve the desired level. Thus drugs in the class known as statins have become the most widely prescribed in the developed world, with good evidence for their ability to lower cholesterol, reduce the risk of heart disease, and prolong life.

Note the similarity here to the kind of information now becoming possible by genetic testing. Cholesterol is not an absolute predictor of the risk of heart disease; there are individuals with high cholesterol levels who do not develop heart disease, and there are individuals with low cholesterol levels who still have heart attacks. A graphic display like those we've already considered for genetic risk factors, Figure 3.6 shows a summary of the effect of total serum cholesterol on lifetime risk of coronary heart disease for men. On the left is shown the risk for men who have a total cholesterol level under 200 (31 percent); on the right, the risk rises to 43 percent for those who have a total cholesterol level between 200 and 239. Standard preventive medicine would strongly recommend reducing cholesterol to a safer zone for those with values over 200. In the context of our RBI equation, R, B, and I are significantly positive for cholesterol testing, and hence screening has entered the mainstream of preventive medicine strategies.

Risk of heart disease
for cholesterol <200:
31 out of 100

Risk of heart disease
for cholesterol 200–239:
43 out of 100

Figure 3.6: Cholesterol is a widely accepted risk factor for heart attack. Note that the increased statistical risk associated with a cholesterol level in the 200–239 range is similar to the diabetes genetic risk depicted in Figure 3.5.

Is diabetes all that different? For T2D based upon the results of the DPP, the evidence for I is actually quite good, and therefore the ability to make predictions about individual risks in relatively young individuals might well be worth incorporating into an individual program of prevention. This argument will only get stronger if the ability to predict R, the actual risk, starts to go up. At the present time, our sampling of R is well short of what it will become in the next few years.

For me, finding out that I carry two copies of the risk allele of *TCF7L2*, and that overall the DNA tests predict a 29 percent risk of diabetes in my future, was not something I expected. My family history is entirely negative for diabetes. But nearly everyone in my immediate family is quite lean—so could it be that we've all avoided this disease by maintaining a normal body weight, despite genetic risk factors? I've gained a bit of weight over the years, and now I'm the heaviest of my siblings. Could I be facing a significant future risk? The DNA test result has forced me to face up to this part of my unhealthy lifestyle. So now I'm undertaking my own mini-version of the DPP: I've embarked on a more disciplined exercise program, I'm paying more attention to what I eat, and I have shed about 15 pounds.

WHERE IS THE MISSING HERITABILITY IN THE GENOME?

The discovery of genetic risk factors for many common diseases in the past couple of years has been enormously exciting. But we are still in the early stages of this revolution. We know, from studies of families and identical twins, that most diseases like diabetes are highly heritable, with, on average, about 50 percent of the risk attributable to genetics. Yet our genetic analysis has so far discovered less than 10 percent of that hereditary component. (Macular degeneration is a notable exception.)

Geneticists all over the world are scratching their heads, wondering where the rest of the heritability is hiding. This has even coined a new phrase: "the dark matter of the genome." Just as cosmologists studying the universe have concluded that the observable part of the universe seems to account for only a small percentage of the total matter that must be present, so those of us studying the genome have to conclude that the genetic risk factors that have been discovered so far represent only a small amount of the DNA variations that must underlie risks for common disease. Where is all the rest?

There are at least four possible explanations:

1. It could, in fact, be the case that most of this resides within a long list of common SNPs that have *very* small effects on disease risk. If the relative risk conferred by a particular glitch is only 1.05, for instance, it might take as many as 10,000 cases and 10,000 controls to be able to discover this particular factor. Most

studies have not been large enough to reach down into this level of subtlety.

2. For many diseases there might be a class of relatively uncommon variants that contribute relatively large effects to disease risk. If each of these was present in less than 5 percent of the population, it would be unlikely to be discovered in most traditional studies, since so few individuals would carry the risk allele.

3. There is another class of common or rare genetic variants that could have more dramatic effects on DNA than the glitches that have been studied in detail: the so-called copy number variants (CNVs), in which a portion of DNA code is repeated many extra times (Figure 3.7). These can be large enough to include one or more genes, yet they could well have escaped detection in most of the disease studies conducted so far. It is certainly plausible that they might lead to a risk of disease. In this regard, exciting new data suggest that newly arising CNVs may well account for some fraction of cases of autism and schizophrenia, although those data remain somewhat controversial.

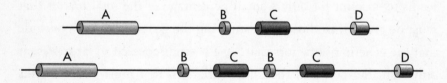

Figure 3.7: A copy number variant (CNV). The individual with the arrangement at the top has just one copy of genes A, B, C, and D. The individual at the bottom has an extra copy of genes B and C. Such CNVs are common in the human population, and may account for some of the "dark matter of the genome."

4. Another possibility, noted previously, is that interactions between genetic risk factors may in some instances be very strong, in which case looking at the individual factors may not fully account for

the overall risk. These so-called gene-gene interactions are being explored by investigators and have not so far yielded much in the way of evidence for large effects. But these are not easy analyses to do, and it is still possible that our assumption of simple multiplication may be wrong.

It seems very likely that at least options 1, 2, and 3 will turn out to be important, and that over the next three to five years a much higher proportion of the heritability for common diseases will emerge. A foreshadowing example is provided by a gene called *PCSK9*. Mutations in the coding region of *PCSK9* were first implicated in cardiovascular disease in rare families that had dominant inheritance of very high levels of cholesterol and cardiac disease.

It turned out that those mutations produced a protein that was overactive, in turn resulting in high serum cholesterol. A creative investigator in Texas, Helen Hobbs, set out to explore the opposite end of the cholesterol spectrum: individuals who have unusually low levels of LDL cholesterol (the "bad" cholesterol). These individuals are protected from heart disease.

It turned out that uncommon variations in *PCSK9*, present in only 1 to 3 percent of the population, accounted for many of those individuals with the lowest cholesterol levels, and reduced the risk of heart disease by as much as 88 percent. In this instance, the variations were actually causing a loss of function of the gene. So here was a yin-yang story: too much *PCSK9* gives heart disease, too little gives protection. This is an example of the second explanation of "dark matter": uncommon mutations causing significant effects.

As the prediction of genetic risk factors gets better and better over the next few years, the argument for empowering healthy individuals with this information in order to practice better prevention is going to get stronger. The ability to predict R in the RBI equation will be steadily improving. The potential interventions, the "I," will also increase, but at a different rate for different diseases. So this is beginning to add up: whether now or in the near future, it's about time to test your own DNA.

DIRECT-TO-CONSUMER (DTC) GENETIC TESTING

In 1996 Donna Shalala, the United States Secretary of Health and Human Services, asked me whether I thought it would ever be possible for individuals to obtain direct analysis of their genome for prediction of future medical risks, without the involvement of a health provider. Secretary Shalala thought of this as a natural next step, in the same way that pregnancy tests had left the confines of the physician's office and found their way onto the pharmacy shelves. I confess that the secretary was more of a visionary than I was then; I found the idea of direct-to-consumer genetic testing completely unimaginable in my lifetime. My, what a difference a dozen years make! As described in the introduction, three companies now offer direct DNA analysis for interested consumers to learn about their risks for a long list of common diseases.

Some of these companies also provide information about risk for traits that clearly are not associated with disease; such information might be referred to as "recreational genomics." Furthermore, the companies also provide information, based on genetic variation, that allows consumers to make predictions about the original geographic location of their ancestors. Appendix E lists the conditions (as of this writing) for which at least one of these companies—23andMe, de-CODE, and Navigenics—claims it can provide information. The appendix also lists the non-disease traits for which information is made available. The cost of the tests ranges from $399 to $2,499.

The introduction of these testing capabilities has created a sensation in the news media. *Time* magazine named one of these companies, 23andMe, the number one invention of the year in 2008, and one of its founders appeared on *Oprah*, describing the concept of DNA testing to millions of people as the show's house doctor (Dr. Mehmet Oz) pronounced his own test results to be highly interesting.

The test is deceptively simple. Since DNA is present in all the cells of the body, it is not even necessary to draw a blood sample in order to initiate the process. DeCODE uses a swab that is capable of scraping enough cells from the cheek to produce sufficient DNA for the study. 23andMe and Navigenics simply have their customers spit into a tube, which is then mailed to the testing company. After the laboratory work is carried out, customers are sent a password so that they can log on to a private Web site and look at their own results. For those who are confused about the answers, Navigenics provides a genetic counselor by telephone, whereas the other two services simply try to give enough information through their Web-based tutorials to help people understand the results.

I described my own experience with DNA testing in the introduction. But has this information actually been useful to others? Well, remember RBI! There are certainly some dramatic anecdotes suggesting that this can provide potential information of considerable impact. One of those was the story of Sergey Brin, with whom we began this chapter, and whose wife, Anne Wojcicki, is a cofounder of 23andMe.

Another is the story of Jeffrey Gulcher. As the chief scientific officer (CSO) of deCODE, he decided to undergo the DNA test that his company was marketing. Jeff was 48 years old and in excellent health.

When he got the results of his genetic test, he was disturbed to see a prediction of a 1.9-fold increased risk of prostate cancer. That result, plus his father's history of prostate cancer at 68, caused him to seek advice from his primary care physician, who suggested that Jeff should have a PSA test, even though that is not normally done until age 50.

Prostate-specific antigen (PSA) measures a substance in the blood that tends to increase in the presence of prostate cancer. There are many false positives and false negatives, and many people are skeptical about the test's value. Jeff's PSA was in the upper range of normal for his age, but a rectal examination showed no lumps in the prostate. If Jeff had not screened his DNA, that would probably have been the end of the story. Yet because of the genetic risk factor, he went to see a urologist. The urologist advised proceeding with prostate biopsy under ultrasound guidance. This can now be done as an outpatient procedure, and is relatively painless. A total of 12 biopsies were obtained from both lobes of Jeffrey's prostate gland.

The results were of great concern. Three of the 12 biopsies, including samples from both lobes, showed prostate cancer. Prostate cancers are graded by their degree of aggressiveness, based upon the appearance of the cells in the biopsy. This so-called Gleason score correlates roughly with the likelihood of metastasis and early death. The Gleason score for Jeff's prostate cancer was 6 (in the middle range).

Debates about the proper management of prostate cancer continue to rage. After all, prostate cancer is generally a very slowly advancing form of malignancy. It is extremely common in older men, and most elderly men with prostate cancer will die with it rather than because of it. However, in a 48-year-old, who might expect to have another 40 or 50 years of life, watchful waiting can be risky. In fact, a recent Swedish study compared watchful waiting with radical prostatectomy in younger men, and demonstrated significantly better survival rates for the surgically treated patients.

Jeff decided to proceed with a radical prostatectomy. The pathological specimen revealed areas of cancer that were even more aggressive than had been discovered by the biopsy, upgrading his disease to Gleason 7. Though this surgery can produce troubling side effects, including incontinence and impotence, Jeff suffered no such complications.

I spoke to Jeff about his experience. I confess that initially I was pretty skeptical. After all, as the CSO of deCODE, Jeff clearly has an interest in selling his genetic test. However, I found him to be reasonable about the conclusions that can be drawn from his own story, quickly admitting that his anecdotal experience is no substitute for much more extensive evaluations of the value of testing. After all, most men with his genetic risk, his level of PSA, and a normal rectal exam would have had normal prostate biopsies. The biopsies would have cost money, and in some instances might even have resulted in complications such as bleeding or infection. Jeff indicated that deCODE is already engaged in a large research study to ask whether genetic testing in combination with PSA is better than PSA alone. My own sense is that stories like Jeff's will mount up, and insurance companies will ultimately see the benefit of paying for testing, to stave off the costs of later treatment for a disease that might have been prevented.

To be sure, these direct-to-consumer (DTC) genetic tests have not arrived without controversy. Debates about whether they are providing empowerment or snake oil have raged in the press. Various medical and policy groups have wrestled with the pros and cons. Concerned that consumers might be misled by the information, some states have moved to block marketing of DTC tests to their citizens, arguing that there is insufficient oversight of the quality of the genetic testing procedure to ensure accuracy. In the United States, genetic test results that are done in-house by the testing laboratory (so-called home brew tests) are not subjected to any real oversight by the FDA.

In the jargon of genetic testing regulation, the labs' only requirement is that they must demonstrate "analytical validity"—basically, that they got the DNA analysis right. But to benefit from having the information, consumers want to know whether the result predicts risk accurately ("clinical validity") and whether the information is actually useful ("clinical utility"). Those latter two criteria are not currently reviewed for home brew tests. The lesson: at the moment you, and only you, can assess your RBI.

A critical question is whether access to this kind of information about risk will actually empower individuals to make changes in their health care behaviors (as it did for Jeff), or whether this will be primarily a recreational experience with no long-term consequences.

Research studies are now getting under way in several medical centers to assess what factors influence an individual's likelihood of incorporating information about genetic risk into their own preventive health care. One such program, the Multiplex Project, is currently offering DNA testing for future risk of illness to 1,000 individuals in Detroit. Each participant is offered testing for eight common adult-onset conditions. The participants are diverse in educational level, ethnicity, and gender, and are followed over the course of several months to see what impact the results have, and what actions they have taken to reduce those risks.

I had the chance to speak with one of the participants about that experience. Lois Klein (not her real name) is 40 years old, and volunteered for the project. She has always been health-conscious, and she thought this was a good chance to learn more about her risks. A blood sample was obtained, and a few weeks later she got the results in the mail. She had a chance to ask questions in a subsequent phone call with a project team member, who confirmed her modestly increased risk of diabetes and colon cancer. Lois shared that news with her primary care doctor, who ordered a glucose tolerance test for diabetes.

That was normal, but Lois was motivated to institute a more regular exercise program, and to incorporate more fruits and vegetables in her diet.

I spoke with Lois a year after the test was done and she indicated that this motivation had persisted. She was aware that she was following diet and exercise routines that she probably should have adhered to anyway, but she found the additional genetic information helpful in inducing a greater sense of urgency to make these changes.

SO PERSONAL GENOMICS IS HERE, BUT CAVEAT EMPTOR

Given the early stage of direct-to-consumer genetic testing, there are experienced and respected individuals in the medical community who are arguing forcefully that it is premature for this kind of information to be made available to consumers. I am not one of them. As someone who has spent the last 25 years in an effort to bring genetics into the mainstream of medicine, it would be odd indeed for me to argue that this information does not belong in the hands of interested parties. The challenge is to be sure the information is valid, and is presented in a way that accurately and understandably depicts what we know and what we do not know.

The American College of Medical Genetics, the primary professional organization of physicians who specialize in this field, disagrees with me. It has recommended that no DTC should be conducted at all, and that such tests should only be ordered by a health care professional.

But the American Society of Human Genetics, the other major organization of genetic professionals, takes a different tack, basically supporting DTC as long as adequate information is available to consumers about the limits of the testing process.

Here is a list of a dozen issues that an interested person might want to consider before scraping the cheek or spitting into the tube:

1. Currently the risk factors detectable by DTC tests are modest in their quantitative contribution to disease. Thus, for most conditions an individual's risk will be changed only slightly by the result of the test. But if you take a test for 20 conditions, it's likely you will have at least one result where your genetic risk is in the top 5 percent of the population.

2. In general, the testing offered does not incorporate an assessment of your family history. As your family medical history is a powerful window into your future, the genetic test can be significantly misleading if it is not considered in that context.

3. The DTC tests generally do *not* detect the less common but highly significant genetic mutations that carry a high risk for disease. For example, these tests do not currently look for all possible mutations in *BRCA1/2*, or Huntington's disease, or fragile X syndrome. If you have a strong family history of a disease, you may wish to seek specific genetic testing for that condition, and you should not depend solely on DTC testing.

4. As noted above, a substantial fraction of the heritability for most common diseases has not been discovered yet. As that additional information comes to light and gets incorporated into DNA analysis, it is inevitable that many individual predictions of risk will require major revision. So if you decide to embark on this adventure of genomic self-exploration now, you should consider it a long-term plan, not a one-shot deal.

5. Although the three companies mentioned in this section pay close attention to quality of the data, you still have to consider the possibility of laboratory mistakes, especially mixed-up samples. Any

consumers would want to have strong evidence that the company they are using has a track record of high quality.

6. The company's interpretation of the DNA test results is not entirely trivial, and different analysts may come up with somewhat different results for the same DNA sample. As described in the introduction, I was tested by all three companies, and there were some significant differences in the report cards that I received. Depending on which variants are actually on the analytic chip the company is using, the results can differ.

7. Most of the current data about risk prediction from DNA testing are based upon studies done on individuals of northern European background. Those cannot necessarily be extrapolated directly to people whose ancestors came from other regions of the world, and to try to do so might result in substantial errors in the prediction.

8. The lack of rigorous information about the intervention (I) factor in the RBI equation can be a major limitation on the utility of the information for many conditions. Consumers who are provided risk information should be skeptical of claims about interventions that might reduce the risk, unless clear references to the utility of those interventions are provided.

9. When interventions are recommended for conditions like diabetes, heart disease, or high blood pressure, they tend to sound obvious. Is it really necessary to pay hundreds of dollars for a DNA test to be told that you should eat a balanced diet, engage in regular exercise, and maintain a normal weight? Of course information about individual risk may be a strong motivator for taking such actions, as was the case for Lois Klein.

10. Be prepared to find that the information provided to you may not be completely transparent, may cause anxiety, and may require you to consult with experts in order to fully understand what you

are being told. The credible companies make an effort to provide information about risk in readily understandable terms, and one of them (Navigenics) even offers telephone consultation with a genetic counselor, but you should be prepared for the fact that you may need some help understanding the information. In that regard, don't count on your primary care physician to be sufficiently well informed about the revolution in personalized medicine to be able to advise you.

11. If you decide to proceed with DNA analysis, you should carefully consider how you want to share this information, and with whom. While recent legislation in the United States has outlawed the discriminatory use of predictive genetic information in health insurance and the workplace (see Chapter 4), there are other possible applications (long-term-care insurance or life insurance, for instance) where an elevated risk of illness or disability can be used against you. The DTC companies assure consumers of privacy protection, but you will need to consider your own plans for sharing the results.

12. The three companies discussed here are conducting their business in a scientifically rigorous way. Yet there is a Wild West of unscrupulous corporations out there, easily found on the Internet. You should be particularly suspicious of sites that offer DNA testing to help optimize dietary intake, and then propose to sell you expensive nutritional supplements to compensate for your defective DNA. The science of "nutrigenomics" is in its infancy. Except for a few well-established examples such as PKU (remember Tracy Beck?), there is relatively little validated information to base such dietary recommendations upon. Some of these companies are running consumer scams.

WHERE IS DTC TESTING GOING?

As the technology for assessing DNA grows in complexity and power, this kind of analysis will extend from an assessment of 1 million variations across the genome to the *complete* DNA sequence of any individual, at a cost of less than $1,000. This is likely to be possible within the next five years. There will be great challenges in interpreting anyone's complete genome sequence, since rare variants seen only in that person will be of uncertain consequence. But it is clear that the genie is out of the bottle, and large amounts of information about the genome will become part of the medical care most of us receive in the not very distant future.

We may soon identify most of the remaining missing factors in heritability (the "dark matter"), making the predictions of risk for future disease more powerful and more accurate. But more than that, we also desperately need more information about the environmental influences on common diseases. After all, since we will not be able to change our genomes anytime soon, many of our hopes for intervention rest upon pinpointing the environmental risk factors, and then modifying those in susceptible individuals. We need better technologies for collecting information about environmental exposure. Though they will be complex and expensive, what we really need are large-scale population studies that follow hundreds of thousands, or even millions, of people prospectively in a way that allows an assessment of how genes and environment interact. Such a study, known as the UK BioBank project, is under way in the United Kingdom; 500,000 people are being assessed by genetic risk factors and medical information. Even there, however, the environmental assessment is relatively modest.

Similar studies are under way in Japan, Germany, and Estonia, and the entire country of Iceland is engaged in such a prospective enterprise, although until now it has been conducted as a private-sector undertaking, and the data are not readily accessible. Oddly enough, in the United States, which traditionally has made the largest investment in biomedical research, there is no current plan for a large-scale prospective study to nail down the genetic and environmental contributions to common disease.

A plan for such a project, the American Genes and Environment Study (AGES), has been in place for five years. It was designed by a distinguished panel of more than 60 investigators who agreed that the value of this study for the future of American public health would be incomparable. But the proposal to enroll at least 500,000 individuals, examine them at least every four years, collect all of their medical records electronically, and conduct a wide variety of laboratory tests including complete genome sequencing led to a cost estimate of about $400 million per year. The government has not yet been willing to fund it.

Yes, this is a lot of money—but it is 0.017 percent of the $2.4 trillion spent on health care in the United States in 2007. This is one of those moments in history, rather like the discussion surrounding the initiation of the Human Genome Project in the late 1980s, when a unique opportunity for medical advances requires leadership in the scientific community and the government.

There is another public policy issue about genetic testing that needs attention. Despite 10 years of deliberation about the need for oversight of consumer genetic testing, relatively little has happened to assure the public that such testing can be trusted. On the positive side, the Federal Trade Commission (FTC) has signaled that it is watching this area closely, and that it will move to shut down companies that are marketing bogus tests. But what is really needed is a coordinated, thoughtful federal oversight process.

An important first step would be the establishment of a public database of information on all genetic tests that are being marketed directly to consumers. Such a database, which most appropriately should be operated by the FDA and should be required of any company that is marketing genetic tests, should include objective information about the predictive value of the test, the populations for which this information is known, the strength of scientific evidence on any claims of benefit, and the possible risks associated with testing. The database should also disclose the certification status of the laboratory conducting the testing, in order for consumers to assess whether there are any concerns about data quality.

Although there seems to be a broad consensus that such a public database would be valuable, at the time of this writing no real steps have been taken to construct it.

CONCLUSION

The era of personalized medicine has arrived. Though important details are still missing, we have already uncovered tremendously valuable information in DNA. Your genome is unique. Do you want to know?

Many individuals are curious about this, and hopeful of improving their chances of a long life. A few of us are already going through extensive DNA testing, and becoming early pioneers for this new paradigm. But many others remain uneasy with such predictions, arguing, "Don't tell me; this will just cause me to worry; I'll deal with disease when it happens, if it happens."

Try a thought experiment. Imagine you already know that you have the following risks: 35 percent chance of diabetes, 20 percent chance of colon cancer, and 38 percent chance of a heart attack. Imagine you have been living with this knowledge for a while. Would it push you into an abyss of depression? Or would it hover in the back of your mind, motivating you to exercise more, eat better, and make sure to schedule that colonoscopy that you've been putting off?

We all have free will. Uncovering the secrets of our DNA will never take that away from us. But it ought to empower us to make better choices.

WHAT YOU CAN DO NOW TO JOIN THE PERSONALIZED MEDICINE REVOLUTION

1. Are you ready to find out your own genetic risks for future illness? To find out more about the testing process, go to each of the sites below and browse through their educational materials. Try out the tutorials. If you decide to go ahead with testing, talk with your close blood relatives to let them know, since what you find out about yourself may have relevance to them too.

   https://www.23andme.com

   http://www.decodeme.com

   http://navigenics.com

2. Obesity is a major risk factor for high blood pressure, diabetes, stroke, heart disease, joint problems, and cancer. Though it is not a perfect measure, obesity is often assessed using the body mass index (BMI), calculated as weight (in kilograms) divided by the square of height (in meters). Do you know your BMI? You can quickly calculate it at http://www.nhlbisupport .com/bmi/. If you have a BMI of 25 or over, it's time to take diet and exercise seriously! See

http://www.nhlbi.nih.gov/health/public/heart/
obesity/lose_wt/control.htm for ideas about how
to get started.

3. Are you at risk for a serious heart attack in the next
    10 years? The long-running Framingham Study in
    Massachusetts has established many of the risk
    factors for coronary artery disease. You can calcu-
    late your own risk, based on age, gender, choles-
    terol levels, and blood pressure at http://hp2010
    .nhlbihin.net/atpiii/calculator.asp?usertype=prof.
    Note, however, that this does *not* include family
    history or any results of genetic testing, so this
    calculation could get a lot better in the future.

CHAPTER FOUR

Getting Personal with the Big C

It was the day before Halloween, 1992. Gathering in the Michigan genetics clinic were: an oncologist, Barbara Weber; a genetic counselor, Barbara Biesecker; a nurse, Kathy Calzone; and me—all about to embark on an adventure into new territory. Our co-adventurers were members of a large family, sitting in our waiting room, about to find out which of them were at high risk of breast and ovarian cancer. This had never been done before, and frankly we were pretty uneasy about how to handle the situation.

The path that led to this unprecedented clinic visit had begun two years earlier, when I attended a late-night unscheduled presentation at the annual meeting of the American Society of Human Genetics. That evening, Dr. Mary-Claire King stunned the audience by presenting evidence of a gene linked to breast cancer. Until that evening, no one had identified this kind of evidence for a highly heritable form of breast cancer, influenced by a single gene. Many in the audience were skeptical. But I was deeply intrigued, especially as my laboratory had just months before identified a gene for a condition called neurofibromatosis in the same region of the same chromosome, 17. I approached Dr. King about a possible collaboration. She agreed, and our laboratories began an intense partnership to try to zero in on the actual gene. Just as with the search for cystic fibrosis, the search for this gene, denoted *BRCA1* for breast cancer gene number 1 (although Dr. King delighted in pointing out that this abbreviation could also stand for Berkeley, California, where her lab was located at the time), was expected to be arduous and frustrating. And indeed it was.

In order to speed up the process of nailing down the precise location of the gene in a sea of millions of DNA base pairs, we needed to identify as many families as possible where multiple individuals had been stricken with early-onset breast cancer. Shortly after Dr. King's announcement, it became clear that individuals in these families were also unusually susceptible to ovarian cancer, so families in which both diseases had occurred were of particular interest.

I enlisted the oncologist and researcher Dr. Weber to join the project, and we began to identify families in Michigan. But they were supposed to be anonymous, and their participation was supposed to be limited to research, not clinical care. It was therefore a truly strange twist to find a large number of members of "family 15" coming to the medical genetics clinic a few months later.

The full story of family 15 is profound. A diagram of the family tree is shown in Figure 4.1. (At the family's request, I have changed the names.) The original contact was through "Dolly," who had been diagnosed with breast cancer at age 48. A decade later, Dolly saw two of her daughters, Janet and Lucy, affected, and ultimately Lucy lost the battle against this terrible disease. Dolly's sister Mattie also developed breast cancer. She had three daughters, two of whom (Pamela and Beth) died of breast cancer before age 30. Pamela also had ovarian cancer. The third daughter, Jessie, fearing a similar fate, had undergone a prophylactic mastectomy.

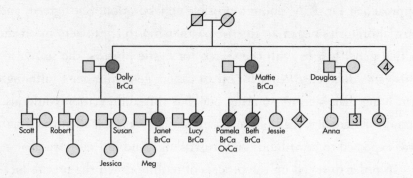

Figure 4.1: The pedigree of the Michigan family, showing many women affected with breast and ovarian cancer. Squares denote males, circles are females.

Dolly had a total of six brothers and sisters. Most of the relatives lived nearby and stayed in close touch. Motivated by a desire to understand what had happened to all these young women, the family followed Dolly's exhortations to participate in our research project, and provided blood samples and access to medical records. We went so far as to track down biopsy samples from deceased members— these samples had been stored in hospital laboratories—so that we could conduct DNA analysis of them as well.

By late summer 1992, it was apparent that family 15 must indeed carry a mutation on chromosome 17, as genetic markers in that region were fully capable of predicting who had been affected with cancer. The chances that this could be a statistical fluke grew smaller and smaller each time a new DNA sample was tested.

Meanwhile, Dolly's remaining unaffected daughter, Susan, was growing increasingly uneasy about her circumstances. She had seen her sister Lucy die from metastatic breast cancer, and her other sister Janet had just been through diagnosis, surgery, and chemotherapy. Susan could not stop thinking about the day when she would find a lump herself. She was also deeply concerned about her 11-year-old daughter, Jessica.

Given her high risk, Susan decided to follow the same track as her cousin, and scheduled a prophylactic mastectomy. Drastic though this procedure was, it seemed to her the best option to try to reduce the risk of the terrible fate that had befallen so many of her family members.

At this point, a remarkable coincidence occurred. Assigned at random to see an oncologist at the University of Michigan, Susan encountered Dr. Weber in the clinic. After hearing her story and her reasons for requesting surgery, Dr. Weber realized that this must be a member of family 15. Just as quickly, she realized that our latest laboratory sleuthing might actually have determined whether or not drastic surgery was warranted. Excusing herself to check the laboratory results, Dr. Weber concluded that Susan had *not* inherited the *BRCA1* risk that had stricken her mother and her two sisters. Her likelihood of developing breast or ovarian cancer was no greater than that for any other woman.

Dr. Weber called me. We had not expected to arrive at this position so quickly in the research project, nor had we anticipated a circumstance like this, in which access to the information would be so urgent. But there seemed to be no choice about the right thing to do. Sitting down with Susan, her affected sister Janet, and Susan's husband, Dr. Weber explained that the results of the research study indicated that almost certainly the planned surgery was unnecessary. Susan was stunned. Later, she recalled feeling as if she were living in a dream. Initially she had trouble believing that this outcome could be possible, but ultimately there were tears of relief all around.

That was late August. As Susan and Janet returned home with their news, other family members quickly picked up on the consequences of this story, inferring correctly that others in the family might also be able to find out their risks. Back at the medical center, we had been preparing for that. We insisted on obtaining new blood samples from all of the family members, so that we could repeat the DNA analysis and make sure that there were no mistakes. The results all checked out.

Initially, we thought about meeting with each family member on a separate day, but it was clear that this large and very close-knit family wished to go through the experience together. And thus it was that on October 30, 1992, about two dozen members of family 15 arrived in the clinic waiting room, as our team went over the plans for our counseling sessions, knowing that there were going to be some surprises.

It was not just the women in the family who came to the clinic. The *BRCA1* risk can be passed down through males. The males tend to have only a slightly increased risk of prostate cancer, pancreatic cancer, and male breast cancer, but can pass the mutation on to their daughters, who then face about an 80 percent chance of breast cancer and a 50 percent chance of ovarian cancer.

As with most families, members of family 15 had not really considered the possibility of paternal inheritance. The family members were brought individually and separately into a counseling room, where a discussion ensued about whether they were sure they wanted to have the results of the testing. Every family member said yes. Then each person's result was presented and the consequences explained. Both of Susan's brothers, Scott and Robert, turned out to carry the *BRCA1* mutation. Both of them had daughters, about whom they were immediately very concerned. Robert demanded that his daughter be immediately tested, and became angry when informed that we did not believe it appropriate to carry out *BRCA1* testing on someone not yet 18 years old.

A poignant session occurred with Janet's daughter Meg, now old enough to receive her results. Janet had hoped and prayed that this familial curse would spare her daughter. But the test was positive.

One of the individuals I counseled provided a different sort of wrenching encounter. This was Jessie, Mattie's daughter, who had undergone a prophylactic mastectomy several years earlier after watching her two sisters die of cancer. The DNA test indicated that Jessie had escaped inheriting the mutation, and that her surgery had been unnecessary. My heart was in my mouth as I shared this result. Yet Jessie took the news with great equanimity, concluding that she had made the best decision that she could at the time of the surgery, and she was comforted because her daughter now need not fear this disease.

Perhaps the most dramatic story, however, was that of Anna. Anna's father, Douglas, was a brother of Dolly and Mattie. While he had watched their struggles with breast cancer with deep concern, he had always assumed that this issue was not relevant to his daughters. Anna had drawn the same conclusion. She attended the clinic that day because she felt it would be supportive to do so for the rest of the family, but she did not expect to learn relevant information about herself.

Nothing could have been farther from the truth. Douglas emerged from his counseling session quite shaken, discovering that he carried the *BRCA1* mutation and, therefore, that each of his 10 children had a 50 percent risk of inheriting it. Seven of those children were daughters, and three of them proved to be carrying the mutation. One of them was Anna. Upon hearing this news, she realized that the breast cancer that had afflicted two of her aunts and four of her cousins now loomed over her as well. At age 39, she had not practiced self-exams, and she had never had a mammogram.

Anxious to pursue this new information immediately, Anna asked whether a mammogram could be done that afternoon. Dr. Weber made the arrangements. The initial reading of the mammogram suggested no abnormality, but after the radiologist learned of Anna's potential high risk, an additional image was taken, revealing a worrisome shadow. A biopsy done a few days later revealed the feared result: Anna had cancer.

Within a few days, Anna had progressed from thinking the family history of cancer was not relevant to her, to a realization of high risk, to a positive genetic test, to a diagnosis of cancer. Recognizing her high risk of breast cancer in the opposite breast as well, she elected bilateral mastectomy, followed by chemotherapy.

Seventeen years have passed since that day in the medical genetics clinic. I spoke to Janet recently to learn what has happened with "family 15." Janet herself continues to do well, but she updated me with many sobering stories about the family. Her youngest brother, Scott, has been diagnosed with esophageal cancer at the age of 43, despite the fact that he is not a smoker or an alcohol abuser. While this kind of cancer has not been statistically associated with *BRCA1* mutations, it is hard to ignore the possible connection.

Janet also had sobering news about her aunt Mattie, who, after surviving breast cancer, was found to have ovarian cancer six years ago, despite having previously had her ovaries removed. Presumably, a small amount of tissue left behind was sufficient to result in this malignancy, which ultimately took her life. To add to the family's tragedies, one of Mattie's sons died of colon cancer at the age of 55, and one of her granddaughters has been diagnosed with breast cancer at age 35.

Janet's daughter Meg, known to be mutation-positive, has so far showed no evidence of malignancy. Nonetheless, she is scheduled to have her ovaries removed in the near future, and potentially a bilateral mastectomy and reconstruction not long after that.

I asked Janet how she copes with all these tragedies. As a person of great inner strength, she simply responded that there wasn't much alternative. She was able to say that despite everything, it has been a blessing to be able to understand the genetic basis for the cancer risk, and to have options available for mutation carriers, even though those options are drastic. In fact, she pointed out, there may well be individuals in the family whose lives have been saved by this knowledge. Anna, for instance, continues to be cancer-free, and may very well have had a tragedy averted by her unexpected early diagnosis.

CANCER IS A DISEASE OF THE GENOME

Many theories about the origins of cancer were raised during the twentieth century, but it was not until the 1980s that the science of molecular genetics began to provide real answers. At that time, much of the focus on cancer was on retroviruses that were capable of causing this disease in other species. The stunning surprise emerging from the evaluation of those viruses was that the cancer genes they carried were actually activated versions of genes present in the genomes of normal animals. This work, which was carried out by Michael Bishop and Harold Varmus, and for which they were awarded the Nobel Prize, demonstrated that our genomes contain particular genes that play an important role in cell growth. But if they mutate in unfortunate ways, these same well-mannered genes can turn bad, leading to unrestrained cell growth and cancer.

Thus, although cancer can strike almost any tissue of the body, giving rise to very different symptoms and consequences, the basic underlying mechanism for all cancers is a disruption of DNA sequence, leading to signals for unrestrained cell growth, like a runaway car with no brakes. These mutated cells grow when they should not, damaging the adjacent tissues. Even more ominously, these cells may enter the circulatory or lymphatic systems, traveling throughout the body and setting up outstations of cancer that all too often lead to death.

The recognition of the role of DNA mutations in cancer, and the pressing need to understand those better, led another Nobel laureate, Renato Dulbecco, to issue the first published call for the complete sequencing of the human genome, in 1986. Dulbecco argued that if we were ever to understand cancer well enough to prevent and treat it effectively, we would need access to the entire human instruction book, both in normal tissues and in cancer cells. Now, more than 20 years later, Dulbecco's dream is coming true.

The story of the new war on cancer is complex and exciting. Many of the changes in strategy and tactics that are now yielding profound new insights are being driven by genomics. Cancer is thus a key area to see how the language of life is revolutionizing our health.

So far, most of the genes that are capable of participating in cancer fall into three categories. The first is the so-called *oncogenes*. These genes code for proteins that normally promote cell growth. Obviously, such genes are critical for development, since we all started as a single cell with a lot of growing to do. These genes are also critical for repair after damage to the body, or in the natural process of cell renewal that is required to maintain health.

Oncogenes are normally tightly regulated, so that their growth signals occur only in appropriate circumstances. A mutation in an oncogene, however, is capable of uncoupling this growth signal from its normal restraints. This is rather like having the accelerator in your car get stuck (Figure 4.2A). For example, the first cancer gene discovered in the human genome is an oncogene called *RAS*. A mutation of just one letter in the coding region for *RAS* is capable of producing a protein that is stuck in the "on" position. Mutated *RAS* is found in a high proportion of colon cancers and bladder cancers.

Cancer genes in the second category are known as *tumor suppressors*. If the oncogenes are the yang, the tumor suppressors are the yin, the brake pedal instead of the accelerator, in that tumor suppressors normally function by slowing down cell growth at times when it needs to be suppressed. But if a mutation inactivates a tumor suppressor gene, then this restraining influence is lost. An important point about tumor suppressors derives from the fact that we humans are diploid. If one copy of a tumor suppressor gene is inactivated, there is still one that is functioning, and so generally the consequences are not that significant. But if the second copy is lost, too, then trouble can ensue. If the front brakes fail but the rear brakes still work, you can still stop before crashing into something. But if you've lost both the front and the rear brakes, you are in trouble (Figure 4.2B).

The most famous tumor suppressor gene is one called p53, sometimes referred to as the "guardian of the genome." Normally, p53 is activated when any damage is done to DNA, in

(A) Activation of an oncogene, acting like a stuck accelerator.

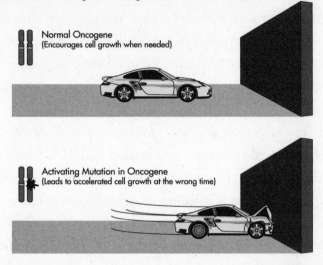

(B) Inactivation of a tumor suppressor gene, resulting in failure of the brakes.

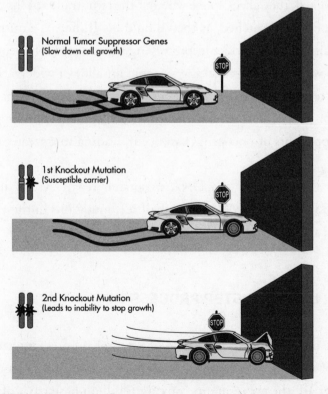

Figure 4.2: Mutations in specific genes can cause cells to grow out of control.

order to stop the process of cell replication until the damage can be repaired. If that is not possible, then the cell commits suicide, thereby preventing the propagation of damaged DNA to subsequent cell generations. The *BRCA1* gene that caused so much havoc in family 15, and in my own family, is also a tumor suppressor gene.

A third type of gene that plays a role in cancer is one that codes for a protein involved in correcting mistakes in DNA. If cancer is a disease of the genome and comes about on the basis of mutations, it is not hard to see that a loss of effective proofreading might increase the risk of cancer. That in fact is the case, particularly for a group of proteins involved in "DNA mismatch repair." These proteins effectively perform like the spell-checker in your word processing program. After DNA is copied, they check to see whether the two strands of the double helix are perfectly matched, as they should be. If there is a mismatch, these repair enzymes correct it, preserving the proper DNA sequence. Problems with the spell-checking system can allow a wide variety of errors to creep into the genome. Of course, most of those will be either negative or neutral in their consequences, but ultimately mutations in tumor suppressors or oncogenes can occur, leading to a greater risk of cancer.

Genes that code for the DNA mismatch repair system include *MLH1, MSH2,* and *MLH3.* As we shall see, mutations in these genes can create a substantially high risk of colon and uterine cancer.

CANCER IS A MULTISTEP PROCESS

If a single mutation in an oncogene or a tumor suppressor gene were capable of causing a progressive malignancy, none of us would be here. After all, the error rate in copying 6 billion base pairs of DNA in 400 trillion cells (the total number in your body) is sufficiently high that such mutations are happening many times each day in each one of us. And yet although one-third of us will ultimately die of cancer, most of us are somehow able to escape this fate. It is increasingly clear that the process of converting a normally well behaved cell to a fully malignant cell requires a number of such mutations acting cumulatively. As shown in Figure 4.3, it is only after a successive accumulation of such mutations that frank malignancy results.

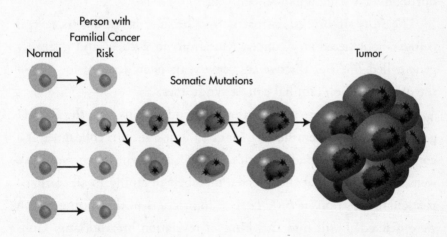

Figure 4.3: Cancer is a multistep process, requiring an accumulation of mutations before a frank malignancy occurs.

This diagram may also make it clear why heredity can increase the risk. Individuals like the carriers of the *BRCA1* mutation in family 15 are already carrying the first mutation step in all the cells of their body, so they are one step closer to trouble. Note, however, that most of the mutations that are present in a cancer cell are not hereditary, but are acquired during the lifetime of that individual. As we shall see, the environment plays an important role as a contributor to acquired mutations. But even in the absence of any environmental insult, the background error rate of DNA copying means that these mutations will be occurring.

BACK TO *BRCA1*

Breast and ovarian cancer require a cascade of events. This explains why not all the *BRCA1* carriers in family 15 fell ill. Some managed to avoid a sufficient number of acquired mutations that have to happen during life for a malignancy to appear.

The story of *BRCA1* is instructive, because it teaches us several things about hereditary cancer. Mutations in *BRCA1* and a similar gene called *BRCA2*, discovered a few years later, account for one of the more common familial cancer syndromes.

Despite the intense work of the King and Collins labs, the *BRCA1* gene was actually identified by Mark Skolnick and his colleagues at a company called Myriad Genetics in 1993. Following their work, we were able to show that the affected members of family 15 all carried a particular mutation in *BRCA1*. Just four letters of the coding region were deleted. I still find this kind of revelation breathtaking. Considering the size of the human genome—3 billion base pairs—it is nothing short of astounding that a change as subtle as deleting four letters is capable of causing such devastating havoc for Janet, Anna, and their affected relatives.

With the identification of the *BRCA1* and *BRCA2* genes, DNA testing to identify specific mutations became possible, and Myriad Genetics lost no time in setting up that kind of laboratory service. However, the circumstances under which this came about continue to generate controversy. Specifically, Myriad Genetics filed for and received a patent on the *BRCA1* and *BRCA2* genes, and this patent gave the company a monopoly on diagnostic uses of these discoveries. Myriad has aggressively protected its patent right by posing legal challenges to any other laboratory that has tried to offer testing. As a result, all competing diagnostic efforts for *BRCA1/2* in the United States have been driven out of business. Myriad thus holds an absolute monopoly with regard to the availability of this test, now sought by many women who have a family history of breast or ovarian cancer.

Although Myriad has offered a test that is highly accurate, the absence of any market competition has kept the price of testing quite high (about $3,500), placing it outside the reach of many individuals who would like to have the information. Third-party payers have often been willing to cover this cost for someone at high risk, but many individuals have been reluctant to ask for insurance coverage, for fear that a positive test might result in a discriminatory action by their health insurer.

Reading this, you may be somewhat incredulous that it is possible for anyone to patent a gene that all of us have as our birthright as human beings. And yet a gold rush of gene patenting occurred during the 1990s, and it is likely that patents have been filed on as many as one-third of all human genes. Many of those patents have now been issued.

The legal argument in favor of patenting is that the gene being patented is not being claimed in its natural state, but is a product of experimental investigation in which the gene has been spliced into a *recombinant DNA vector*, sequenced, and analyzed. The United States Patent and Trademark Office has chosen to accept the argument, based on an analogy with chemical patents, that this is "composition of matter" appropriate for patent protection. In effect, the government has chosen to use patents to reward people for the effort of discovery. That position has recently been challenged by the American Civil Liberties Union, which filed suit seeking to invalidate the Myriad patents.

Intense disagreements have raged over the last 15 years about the appropriate role of patenting in the human genome. In trying to assess the arguments, it is well to recall that the original purpose of patent laws was not simply to enrich inventors, but to provide an incentive for an inventor who had discovered something of potential public benefit to make the investment necessary to turn that discovery into a useful, marketable product. The availability of a patent provided an assurance of a monopoly on the market for a limited time (17 to 20 years), which might then allow the inventor to recoup the developmental investment, and even to make a profit.

Yet it is not exactly clear how this argument applies to patents on human genes. Perhaps the discovery of a gene that leads directly to a promising pathway for treatment of a disease might fit the concept, since the long road from a basic scientific discovery to an approved therapeutic agent takes many years and requires an investment of hundreds of millions of dollars. In the absence of patent protection, biotechnology or pharmaceutical companies would be unlikely to pursue such therapeutic development.

On the other hand, I think this argument falls flat when it comes to diagnostic applications. The technology for sequencing or genotyping DNA to look for specific mutations is already highly evolved, and will continue to develop even greater cost efficiency and accuracy in the future. Thus, the supposed need to provide an incentive for companies to develop DNA diagnostics is unconvincing. In that situation, many of us would argue that it would be better for the public to have competition in the marketplace, in order to provide an incentive for higher quality and lower price.

Recognizing these principles, my own laboratory and that of Lap-Chee Tsui insisted that the discovery of the CF gene, in 1989, be available on a nonexclusive basis to any laboratory that was interested in offering testing. Many legal scholars have pointed to this as a better example of how to ensure public benefit.

Furthermore, I donated all my own patent royalties from the CF gene discovery to the Cystic Fibrosis Foundation, to support further research into treatment. As director of the Human Genome Project, I took further steps to discourage unwarranted gene patenting, insisting that all information about the human DNA sequence be placed immediately in the public domain. The information contained in our shared instruction book is so fundamental, and requires so much further research to understand its utility, that patenting it at the earliest stage is like putting up a whole lot of unnecessary toll booths on the road to discovery.

WOULD YOU WANT TO KNOW ABOUT A CANCER RISK?

The wide availability of *BRCA1/2* testing has led many women with a family history of breast or ovarian cancer to seek this information. When we met with "family 15" in 1992, we were not at all sure what to recommend to those who carried the mutations, but since that time much information has been derived by carefully planned research studies. Recommendations and conclusions continue to evolve as new data appear, and anyone in these circumstances needs to consult regularly with an expert; nevertheless, some of the conclusions can be summarized as follows:

- Women with mutations in *BRCA1* or *BRCA2* should be counseled about their high risk of breast and ovarian cancer; and the various surgical and nonsurgical options should be clearly presented, with ample time for discussion and for answering of questions.
- A woman who chooses watchful waiting would be well advised to consider periodic MRI scanning. Mammograms alone, which

are appropriate in other circumstances, may not be sufficient to detect early signs of cancer in these women who are very high risk.

- When the risk involves ovarian cancer, watchful waiting is considerably less reliable. A blood test detecting a substance called CA–125 can indicate the presence of ovarian cancer, but often only after the cancer has spread. Similarly, transvaginal ultrasound studies of the ovaries have been advocated, but there is only modest evidence that they actually detect ovarian cancer early enough to be useful.

- Prophylactic surgical removal of the ovaries at the completion of childbearing is an option that should be seriously considered, since this reduces the risk of ovarian cancer almost to zero. It is important that the Fallopian tubes also be removed, since these can also be the site of cancer. This surgical procedure will induce menopause, but that can be hormonally managed in most instances.

- Perhaps the most difficult decision relates to prophylactic mastectomy. If properly performed, this can reduce the future risk of breast cancer almost to zero. Many women understandably recoil at the drastic nature of this procedure. All women in these circumstances should have the opportunity, however, to learn about reconstructive procedures, which can now be carried out at the same time as the original surgery, and which many women find to give a satisfactory cosmetic result.

- Males with *BRCA1* or *BRCA2* mutations face a much lower lifetime risk of cancer but do have a modestly elevated chance of developing prostate and pancreatic cancer. Males with *BRCA2* mutations are also at risk of male breast cancer, and should be counseled accordingly about careful surveillance.

- No one should undergo a DNA test for cancer susceptibility—or, for that matter, for susceptibility to any other disease—without having a chance to learn about all the possible consequences. The burden to provide this kind of pretest education falls heavily upon direct-to-consumer testing companies, and they have been variably successful in meeting the challenge thus far. Education must include the psychological impact of finding that one is at increased risk for a serious illness. Most studies indicate that after a period of a few months of adjusting to bad news, most individuals return to their baseline sense of well-being, even for conditions in which the risk is very high, such as *BRCA1*, Alzheimer's disease, or Huntington's disease. But even a test that comes back negative may have problematic psychological consequences; for instance, Susan experienced a prolonged period of "survivor guilt" after testing negative for *BRCA1*, when so many other family members were affected.

ADDING DISCRIMINATORY INSULT TO GENETIC INJURY?

A major concern for many individuals in the past has been the risk of genetic discrimination, especially in health insurance or employment. The fear that a test indicating a heightened susceptibility to future illness might result in such discrimination has caused many individuals to forgo genetic testing. Others have decided to pay for testing out of their own pocket, or to use an assumed name, or both, so that the information would not find its way back to their medical records. A particularly poignant story from a few years ago illustrates just how destructive this situation can be for the health of the individual.

A woman physician I know, living in Chicago, had a strong family history of breast and ovarian cancer, and was of Ashkenazi Jewish background. Knowing that the risk of a *BRCA1* mutation is somewhat higher in this population than the general public, she arranged to have testing carried out privately, paying for it herself. When the test came back showing she carried a mutation in *BRCA1*, she informed her primary care physician, but asked that the result not be recorded in her medical record. Though the 1996 Health Insurance Portability and Accountability Act (HIPAA) had outlawed genetic discrimination for group health insurance policies, this woman supposed that she might at some point require an individual policy, and she did not want to risk becoming uninsurable.

A year or so later, she experienced some lower abdominal pain. Given her anxiety about the possibility of ovarian cancer, she consulted her physician, who ordered an ultrasound study. The radiologist, unaware of her *BRCA1* mutation, assumed this was a routine study and read it as normal. The discomfort faded away, and the woman convinced herself that she was just being overly sensitive to normal aches and pains. But a year later the pain returned, and this time it was more sustained. A repeat ultrasound was obtained. To the shock of the woman and her doctor, it showed stage 4 ovarian cancer. A retrospective review of the initial ultrasound, now informed by her high-risk situation, revealed subtle findings in one ovary that might well have led to a diagnosis a year earlier, while the cancer was still contained. This represents a true tragedy, in which a life may have been shortened because of fear of discrimination.

The need to address the problem of genetic discrimination had been pointed out as far back as 1990. Working together with the advocacy community, particularly breast cancer support groups interested in *BRCA1/2* testing, a group of us issued a call for federal legislation to prevent genetic discrimination in health insurance and in the workplace. Representative Louise Slaughter of New York responded, introducing the first legislation on this matter in 1996. The principle seemed blindingly obvious: since none of us can choose our

DNA, it should not be used to discriminate against us—any more than we should be discriminated against because of the color of our skin.

But sequencing the human genome proved to be substantially easier than getting a well-crafted bill passed by both houses of Congress and signed into law. Objections from the powerful health insurance industry and various employers' organizations (most prominently the Chamber of Commerce) kept the various bills from making much progress until April 25, 2007, when a bill passed the House of Representatives. Poetically and completely coincidentally, April 25 has traditionally been the date when schools in the United States have celebrated "DNA Day," since this corresponds to the date of the publication of Watson and Crick's description of the double helix in 1953. What a DNA Day gift this was!

It wasn't until April 24, 2008 (one day short of the next DNA Day!), that a related bill passed the Senate, and soon thereafter the bill passed the House.

On May 14, 2008, President Bush signed the legislation, as a few of us gathered around him in the Oval Office. It had taken 12 years to achieve this victory. Had that legislation been in place several years earlier, my friend in Chicago might have been spared the devastating consequences of stage 4 ovarian cancer.

OTHER HEREDITARY CANCERS

Hereditary breast cancer and ovarian cancer have attracted a great deal of public attention, because of the dramatic consequences of living in an affected family. But many other cancers can also be highly heritable, and a strong family history of cancer, especially cancer that comes on at a relatively early age, should be an impetus for further investigation. On the list of such highly heritable malignancies are tumors of the endocrine system, known as multiple endocrine neoplasia; childhood tumors of the eye called retinoblastoma; hereditary tumors of the kidney, pancreas, and nervous system called von Hippel Lindau syndrome; and a relatively common condition, neurofibromatosis (NF1).

Neurofibromatosis affects about one in 3,000 individuals and is inherited as a dominant trait. Yet the relevant gene has a high mutation rate, so that new cases often appear without any family history. The disease is characterized by benign flat brown spots on the skin— "café au lait" spots—as well as the appearance of characteristic fleshy tumors of the skin, called neurofibromas, often at the time of puberty. These can become extremely numerous and cosmetically disfiguring. Furthermore, occasional tumors can be large, can grow rapidly, and can compromise bodily function. There is also a risk of learning disabilities and cancer, especially tumors on the nerve to the eye. My own laboratory identified the gene for NF1 in 1990, and ongoing research is aimed at using the information about the gene's normal function to develop an effective targeted therapy. Thus far, however, those efforts have not resulted in a major advance.

After breast and ovarian cancer, perhaps the most important hereditary cancer is that of the colon. Unlike many other body sites, the colon is readily accessible to examination by modern flexible colonoscopy. Furthermore, colon cancer follows a stereotyped progression from a benign polyp to a localized cancer to an invasive cancer. This process takes many years, and thus intense medical surveillance and removal of polyps can be lifesaving for those at high risk.

In one particular dominant condition, familial adenomatous polyposis (FAP), the colon is filled with hundreds or thousands of such polyps, coming on as early as childhood. This makes it an impossible task to remove individual polyps, and so the treatment requires complete removal of the colon (colectomy). Failure to recognize this condition is a tragedy, as it invariably leads to cancer, usually by age 40. Of course, no one would choose to have a colectomy if it were not necessary, but the surgery can be done so as to preserve the patient's chance of leading an essentially normal life, and any individual in a family with FAP who is found to be affected would be very ill advised to forgo it.

Whereas FAP is relatively rare, a much more common condition, and a critical one to recognize, is another form of highly heritable colon cancer associated with only a small number of polyps, but with very high potential for malignancy. Variably known as Lynch syndrome—after Dr. Henry Lynch, who first described it—or more descriptively but less pronounceably as hereditary nonpolyposis colon cancer (HNPCC), this condition is also inherited as a dominant trait, and it leads to an approximately 60 percent risk of colon cancer and a 30 percent chance of uterine cancer in individuals who carry the mutation.

A fascinating detective story led to the identification of the genetic cause of this condition, namely mutations in *MLH1, MSH2,* or *MLH3*. These genes code for proteins that are part of the spell-checker for the genome. As these genes are expressed in all tissues of the body, it remains something of a mystery why the major risks of mutation seem to be in the colon and the uterus, but presumably there must be other, compensatory factors in other places.

Careful study of HNPCC has led to very strong recommendations about how to identify families that may harbor one of these mutations, and what interventions to offer those found to carry the very high genetic risk. Specifically, individuals with one or more first-degree relatives with colon cancer at age 55 years or younger should consider having the family evaluated. Genetic testing for *MLH1, MSH2,* and *MLH3* mutations can then be conducted, most appropriately on an individual who has been diagnosed with cancer. Once a mutation has been discovered, then other family members who are interested in the information can be tested for that specific misspelling. Those found to be positive should begin colonoscopy at age 25 or 30, and should have it faithfully every year, not at the longer intervals recommended for the general public (starting at age 50 and redoing the screening every 5 to 10 years). Women with HNPCC mutations should also have regular endometrial samplings, and should consider hysterectomy after childbearing in order to avoid the risk of uterine cancer.

Recognition of this hereditary cancer, and the establishment of an effective means for preventing its previously devastating outcome, has already saved the lives of many people. One of them is Jim Green.

Jim (not his real name) remembers when HNPCC first became a reality in his family. He felt a sense of foreboding as soon as he heard the voice on the other end of the phone. His brother Steve, always a happy-go-lucky guy, sounded deadly serious this time. "Jim," he said, "I'm calling to tell you I have been diagnosed with colon cancer. And because I'm only 32 years old, and there seem to be others in our family who have had cancer at an early age, this might also be relevant for you."

Jim was stunned. He was aware that his grandmother had had cancers of some sort, but the family had not discussed them much in the past. How could his brother, a navy officer and the picture of health, develop cancer at such an early age?

Over a few months, the facts began to emerge. Jim and Steve's grandmother had indeed had uterine cancer, but she had subsequently succumbed to colon cancer. Two of her siblings had also died of cancer, as had her mother. Yet Jim and Steve's mother, who represented the genetic connection in this family, appeared to be just fine.

Steve was determined to get to the bottom of this, and had his colon cancer tested to see if there might be clues as to its cause. Eventually, he was seen both at the Mayo Clinic and at Johns Hopkins, where a DNA analysis was undertaken, and he was found to have a hereditary deletion in the *MSH2* gene. That meant the family was at risk, and the finding of a specific mutation made it possible to investigate others.

Jim decided to be tested. Somehow he always knew the test would be positive, and it was. But he was determined to be proactive, and so he began annual colonoscopies almost immediately. The first two or three tests were normal. Then, a year ago, two polyps were found and removed. These might have gone on to become a malignancy, just as had happened with his brother. Fortunately, Jim is in a health care system where this procedure is largely covered by insurance.

Jim's major concern now is about his children. His two sons and a daughter are all still under the age of 10, but they are beginning to ask questions about why Daddy has to have this procedure done every year. He is aware that fairly soon he will need to explain to them that this condition might also be relevant for them, though to protect their opportunity to make their own decisions about genetic testing, this would ordinarily not be offered until they reach the age of 18.

This family story has been a profoundly unsettling experience for Jim. Yet he now says it drew the family closer, and he and his brother are both doing well.

WEAKER GENETIC RISK FACTORS FOR CANCER

Most cancers are unlike the examples discussed here, in which a single mutation plays a key role. In fact, individuals with mutations in *BRCA1/2* or the HNPCC genes account for only 5 to 10 percent of breast and colon cancers respectively. But the other 90 to 95 percent are not completely free of hereditary influences. In fact, very recent discoveries, using the genome-wide association strategy we discussed in Chapter 3, have identified a much longer list of relatively common susceptibility variants in an entirely different set of genes, each of which confers an increased risk of a particular type of cancer. Such discoveries have already been made for breast cancer, prostate cancer, and colon cancer, and others are coming along quickly.

Although these more recently described genetic risk factors are quantitatively weaker, they are carried by a larger number of people, so their impact on cancer overall may be substantial. You met one example in the person of Jeff Gulcher, who was led to discover a potentially life-threatening prostate cancer after testing for these risk factors. There is less information, however, about exactly how individuals found to be at modestly increased risk from these newly discovered gene variants should modify their health-related behavior in order to reduce that risk, and this is now a major research topic.

CANCER IS A DISEASE OF THE GENOME, BUT MOST CANCER MUTATIONS ARE NOT HEREDITARY

As shown in Figure 4.3, many steps are required for a normal cell to reach a fully malignant state. The vast majority of those mutations occur after birth, at some point during an individual's life. Each successive mutation that affects cell growth, whether in an oncogene, a tumor suppressor, or a DNA mismatch repair gene, predisposes that cell to grow a little faster than its neighbors. In a process that has some similarity to "survival of the fittest" in biological evolution, those mutations accumulate, and rare cells with additional mutations gain growth advantages over their neighbors. Fortunately, the body's immune system and other defenses recognize most of these early cancers before they get very far, and stop them. Only those cancer cells that manage to evade surveillance can become a threat to life.

Where do these mutations come from? It is tempting to assume that they must all be caused by some external influence, and to imagine that in a completely natural state, humanity would be free of such glitches. However, this is almost certainly not correct: random mistakes in the process of copying DNA are simply a part of life. In fact, it is remarkable how little biological mayhem we face, despite the 6.2 billion DNA letters that have to be copied each time a cell divides. Most of your cells have already gone through dozens of copying steps since your conception.

It would be a mistake, however, to assume that all these mutations are just due to bad luck. Another major influence comes from the environment. Of all the environmental contributions to cancer, smoking ranks number one as the most dangerous. Smoking directly alters the DNA in the mouth, esophagus, and lungs, and it also raises the risk of cancer in the pancreas, bladder, colon, and elsewhere. It is a mutation-maker par excellence. The evidence for the elevation in the risk of lung cancer is overwhelming: approximately 87 percent of all cases of lung cancer are directly attributable to cigarettes. Deaths from cigarette-induced lung cancer in the United States are equivalent to the crash of a jumbo jet every day of the year, and that doesn't even account for all of the deaths from emphysema and heart disease. Most cancers of the larynx and oral cavity are also caused by cigarettes. On the average, smokers lose 12 years of life. Cigarette smoke contains not just one but many compounds that damage DNA, leading to mutations. These compounds, known as carcinogens, are also present in so-called smokeless tobacco, and thus produce a high risk of oral cancer in those who chew tobacco or dip snuff.

For all but a few individuals, it is extremely unlikely that any genetic factors will ever be found that confer a risk of cancer anywhere near as high as smoking cigarettes. This is an entirely preventable risk, but the challenge of eliminating it from our population is clearly enormous. Young people, attracted to the dangerous habit of smoking by peer pressure, role modeling in films, and widespread advertising, rapidly become addicted to the nicotine in cigarettes. The combination of this physical addiction and a youthful sense of immortality leads to a general lack of interest in quitting smoking until somewhat later in life. At that point, the addiction is full-blown, and stopping may be quite difficult. But even after decades of heavy smoking, quitting can be enormously beneficial. Within 10 to 15 years after quitting, the risk of heart disease is the same for a former smoker as for someone who has never smoked, and the risk of lung cancer has been cut in half.

If you are a smoker, there is nothing more important that you can do for your own medical future than to stop. At the end of this chapter are listed some resources that have helped 30 million Americans achieve this goal. You can be one of them!

Ultraviolet radiation, absorbed by the skin on a sunny day in the absence of sunblock, damages DNA in exposed areas in a very specific way. The effects of such exposure are particularly dangerous in children, because of the future risks of melanoma, a particularly malignant type of skin cancer. Pediatricians now strongly recommend against sun exposure sufficient to cause sunburn or even significant tanning in children.

Cosmic rays that bombard us every day from outer space may account for some of the mutations that we all accumulate. Radiation used for medical diagnostic purposes can also damage DNA, but the dose is generally limited to the lowest possible level, at which the risk is thought to be very slight. Therapeutic radiation for the treatment of cancer can have the paradoxical result of killing the cancer cells acutely, but leading to a heightened risk of an independent secondary tumor many years later.

Although there is much concern about industrial chemicals, and the role they may play in the high incidence of cancer in many parts of the world, it has generally been difficult to pin down specific culprits. One exception is the organic compound benzene, which can be activated by the body into carcinogenic compounds and can increase the risk of leukemia. Others are asbestos and radioactive compounds such as uranium.

Even compounds naturally produced by other living organisms can be carcinogenic. A dramatic example is a toxin produced by a particular fungus, aflatoxin, which is a potent inducer of liver cancer. This toxin has sometimes turned up in foodstuffs such as peanuts, where long-term storage has resulted in fungal contamination.

Other dietary factors in cancer remain a topic of considerable controversy. Some experts passionately advocate specific diets to reduce the risk of cancer. It is clear from long-term studies of large numbers of individuals that a diet high in fruits and vegetables, and low in red meat, is associated with a reduced risk of cancer. But the precise nature of this protective effect remains elusive.

In Asia, the high risk of gastric cancer has traditionally been blamed upon diet, particularly on the method of cooking fish. More recently, it seems that a bacterial organism, *Helicobacter pylori*, capable of living in the stomach, may be the major culprit in Asian gastric cancer. (it also has a better-established role in causing ulcers).

Other major contributors to cancer risk, of special importance because they are preventable, include a certain set of oncogenic viruses. The realization that most cases of cancer of the cervix in women are attributable to exposure to a particular virus—human papillomavirus (HPV)—and that this virus is nearly always acquired by sexual contact, has revolutionized our understanding and resulted in a massive public health effort to prevent this life-threatening disease by use of a vaccine. Clinical trials have demonstrated a high level of effectiveness for the vaccine, which to be most effective needs to be given to girls prior to the onset of sexual activity. A strong argument can also be made that boys should be vaccinated, since males can carry and spread the virus, even though they do not suffer the same high risk of cancer. Other viruses strongly associated with cancer for which the preventions and interventions are not yet as effective are hepatitis C and liver cancer, and Epstein-Barr virus and cancer of the head and neck, especially in Asia.

TOWARD A COMPREHENSIVE UNDERSTANDING OF CANCER

Cancer treatment has come a long way in the last 50 years. To cite just a couple of examples, childhood acute leukemia, once nearly always fatal, is now cured by aggressive chemotherapy 85 to 90 percent of the time, though the therapy is admittedly quite toxic. Similarly, Hodgkin's disease, a particular form of lymphoma that often strikes young people, is now almost always cured, even if it has spread to multiple lymph nodes throughout the body. And who among us cannot be moved and inspired by the story of Lance Armstrong, seven-time winner of the Tour de France, who has been completely cured of a highly malignant form of testicular cancer that had spread to many parts of his body, including his brain?

Yet along with all those dramatic successes, there are many heartbreaking stories of surgery, radiation, or chemotherapy that failed to provide the hoped-for cure. It is still the case, nearly 40 years into the "war on cancer," that most of the treatments we use represent blunt instruments. Yes, they attack the cancer cell by targeting its rapid growth, but they inflict serious collateral damage to other, normally growing cells, particularly in the bone marrow and the gastrointestinal tract. Our cancer therapy is all too often like carpet bombing.

Instead, we need smart bombs. But a bomb can be smart only if you know the precise target. Until recently, we have had an inadequate understanding of what those ideal targets might be.

GOING BEHIND CANCER'S FRONT LINES

The same tools that led to the success of the Human Genome Project are now being applied to determine the complete DNA sequence of many cancer cells. If cancer is a disease of the genome, the instrument that we need most to reveal its secrets is a highly efficient, highly accurate, low-cost DNA sequencer. That instrument is now at hand.

Even prior to the advent of high-throughput DNA sequencing, about 300 genes had been identified that appeared to carry significant mutations in one or more types of cancer. But these discoveries were generally arrived at by looking at a small subset of all the genes in the genome. An exciting example of a much more comprehensive approach is a project called The Cancer Genome Atlas (TCGA), which has recently been initiated as a joint effort of the National Cancer Institute and the institute that I previously led, the National Human Genome Research Institute. This was mounted initially as a pilot project focused on three malignancies: brain tumors, ovarian cancer, and lung cancer. The goal was to apply all the new, powerful tools of genomics to catalog all the mutations in several hundred cancers of each type. The first pilot data from TCGA, focused on the highly malignant brain tumor called glioblastoma multiforme, revealed a number of stunning surprises, including identification of several genes not previously known to play a role in this tumor. Ironically, one of those genes was *NF1*, which despite its discovery almost 20 years ago in my own laboratory, had not previously been implicated in brain cancer. Another gene, *ERBB2*, showed the type of mutations that suggest that certain types of targeted therapy not previously tried for brain cancer might actually be quite effective in a subset of patients.

At about the same time, a study of almost 200 lung cancers of the type specifically referred to as adenocarcinoma revealed several new targets not previously identified, despite decades of prior research on this type of cancer. Additional studies on breast cancer, colon cancer, and pancreatic cancer are revealing similar surprises, and this deluge of discovery is likely to accelerate over the next few years.

The first complete sequencing of a cancer genome, not just focused on a selected subset of genes but covering the entire genome, has just been published for a particular type of leukemia, and has revealed some major new insights.

This detailed approach is filling out our picture of the steps involved in the transformation from a normal cell to a fully malignant one, demonstrating a significant accumulation of mutations. The evidence also suggests that cancers from different people that we used to consider as homogeneous because of their appearance under the microscope may be profoundly different at the DNA level, with marked consequences for prognosis and response to therapy. Some of these revelations are already finding their way into clinical practice, as in the case of Karen Vance in Chapter 1, where analysis of gene expression in her breast cancer made it possible to predict its low likelihood of recurrence, and spared chemotherapy that would have been both toxic and unnecessary.

FROM MOLECULAR DISCOVERY TO MAGIC BULLET

There is no question that our basic understanding of cancer has made profound progress in the last 25 years, since the discovery of the first oncogene made it clear that cancer is a disease of DNA. It is the fervent hope of patients, providers, and researchers alike that these new insights will lead as rapidly as possible to highly specific therapies that are both effective and nontoxic. That may sound like science fiction, but the dream is starting to come true in a few instances, and promises to become the dominant approach to the production of an entirely new generation of cancer therapeutics. For a glimpse of what this promise brings, consider Judy's story.

Judy Orem was diagnosed with chronic myeloid leukemia (CML) in 1995, when a blood test showed that her white blood cell count was 66,000, about 10 times normal. Her grandmother had died from the same disease, and her mother suffered from a different type of leukemia, so Judy knew that this was very serious. She was initially treated with interferon, which made her feel as if she had a bad case of the flu, but despite this chemotherapy and another drug added a year later, her disease progressed steadily.

By the fall of 1998, she was told that she had only months to live. She arranged a trip to New Zealand with her family. Judy specifically planned this to be a farewell trip, to try to create good memories for her family as she faced the end of her life. At that same time, she met Dr. Brian Druker, whom she had heard about through the Leukemia and Lymphoma Society. He told her about a new experimental drug, Gleevec, and said that she qualified for a clinical trial—but she was required to stop all other medicines first.

Returning from the trip, Judy got an apartment in Lake Oswego, Oregon, in order to be close to the site of the clinical trial. She was only the ninth person ever to receive Gleevec. Judy started a support group and produced a newsletter for other patients involved in the trial, and they met regularly with Dr. Druker. The support group meetings turned out to be a useful format for addressing some of the problems that some participants were encountering, such as nausea. That problem was cleared up if the medicine was taken with food.

Over the next few months, Judy began to feel progressively stronger. She was thrilled and amazed to learn that after just five months, her white count had fallen to normal and that the chromosomal marker of her leukemia cells, the so-called Philadelphia chromosome, had dropped to 5 percent of the cells in her bone marrow.

No one could predict the durability of the response, but as the months went by Judy found her life returning to normal. Every year seemed like an amazing gift. Now, 10 years after the initiation of the treatment, her Philadelphia chromosome is undetectable, though a very sensitive molecular test can still find a faint trace of the malignant cells. She expects to remain on the drug indefinitely. When I spoke to her she was happily babysitting for her two grandchildren. "I never dreamed I would know them at all," she said, "and now every day as a grandmother is a blessing."

I asked her what she would want people to know about these developments in the treatment of leukemia. Without hesitation, she said, "There is real hope out there."

What was the secret of the drug that brought Judy back from death's door? In this case, the story stretches back over several decades. In the 1960s, Dr. Janet Rowley and colleagues used methods to visualize human chromosomes to show that the vast majority of individuals with Judy's disease (CML) have a characteristic rearrangement of their chromosomes. Specifically, in the leukemia cells, parts of chromosomes 9 and 22 are joined together to make a small derivative chromosome—this is the "Philadelphia chromosome" mentioned above and named for the city of residence of the researchers who first described it.

When molecular biology came along, it was possible to demonstrate that the translocation of those two chromosomes occurred in a very specific way in virtually all patients with CML, connecting part of a gene on chromosome 9 called *BCR* to part of another gene on chromosome 22 called *ABL*. The result was a gene formed by the fusion of two others—a chimera—capable of coding for a chimeric protein. This protein, which does not occur in any normal cells, was apparently a bad actor, capable of causing a normal white blood cell to grow out of control. This resulted in the clinical phenomenon we call leukemia (which literally means "white blood"). The unrestrained proliferation of malignant white blood cells crowds out the other elements of the bone marrow, infiltrates other organs, and leads to death if not stopped. Since this chimeric protein does not occur in normal cells, it is an ideal target for drug development. Any compound that blocks this protein would be expected to be effective against CML, with few if any side effects.

An interesting partnership was initiated between a university investigator, Dr. Brian Druker, and a pharmaceutical company, Novartis, to see whether small molecule compounds, derived in a fashion similar to that described for cystic fibrosis in Chapter 2 and Appendix D, could be identified that would block the action of this protein. Druker and Novartis identified a long list of candidate compounds. One of these compounds appeared to be quite effective in cell-culture tests, and even seemed to work in mice that had been engineered to develop CML. But all this indirect evidence was insufficient to make an accurate prediction of how the drug would work in human patients.

In the initial trial for Gleevec, 32 patients who were suffering from far-advanced CML, with a very limited predicted life span, were given oral doses of the drug. Judy was one of them. To the surprise and delight of everyone involved, 31 of those patients (including Judy) showed an immediate drop in their white blood cell count, and achieved a complete remission within a few weeks. Most of them remain in remission today. In fact, a recent prospective study of patients with CML treated with Gleevec indicates that 95 percent of them can expect to achieve a remission that lasts at least five years, though they need to remain on the drug. The drug appears to have relatively few side effects, can be taken as an oral pill once a day, and is very well tolerated.

There is, however, a major issue with Gleevec. With no other competition in the field, and a desire on the part of Novartis to recoup its research investment, patients taking Gleevec must spend approximately $40,000 a year to stay on the therapy. Third-party payers will generally cover this, but more than 46 million U.S. citizens have no health insurance, and that situation presents a major challenge for offering this lifesaving therapy to many who need it.

Dr. Druker happens to be Judy's doctor, but this therapy is now available from any oncologist, having been approved by the FDA as first-line therapy for the treatment of CML.

Interestingly, Gleevec is beginning to find other uses as well. Apparently the chimeric protein is present only in CML, but related proteins are activated in other tumor types and have binding sites that are very similar in shape.

The case of my friend Marvin Frazier provides a dramatic example of how the molecular relationships between cancers that appear to have almost nothing in common can turn out to be lifesaving.

It was 1998, and the leaders of the International Human Genome Project had gathered in Bermuda to assess early progress in the audacious goal of reading out all the letters of the human DNA instruction book. As the leader of the U.S. effort, I needed to keep my mind completely focused on the matters at hand. Yet one part of me was deeply shaken and distressed that day by some news that spread through our group: our missing colleague Marvin Frazier, a significant leader in the Department of Energy's genome program, had just been diagnosed with a massive abdominal tumor. The tumor had already spread to his liver and was expected to progress very rapidly. We all wrote heartfelt notes to Marvin and assumed that we would never see him at a scientific meeting again.

Over the next two years Marvin went through four major abdominal surgeries to remove portions of this bulky tumor, a gastrointestinal stromal tumor (GIST). But despite surgery and intensive chemotherapy, the cancer was inexorably progressive. Two years after his diagnosis, Marvin was ready for hospice, taking high doses of narcotics for intractable pain and preparing himself and his family for the end. But as Marvin searched the Internet, he discovered new information that at least some cases of GIST were associated with activation of the oncogene *KIT.* Marvin knew that the product of this gene belonged to the same class of molecules as the protein that was the target for Gleevec. So there was some chance that Gleevec might also help him, though this seemed like a long shot.

Marvin enrolled in a very early clinical trial, though his severe end-stage cancer made it seem quite unlikely that any real benefit could be achieved. Within just one week, his pain resolved to the point where he was able to discontinue narcotics. Within a month a scan showed that his tumors had shrunk by 50 percent. As time went on, the tumors became less and less perceptible. After a few more weeks, Marvin went back to work. I'm happy to say I've had many occasions since that time to talk about science with Marvin. He is aware that his cancer is probably not cured. In fact, seven years after first starting Gleevec he has had some regrowth of tumors, requiring a higher dose of the drug, and the addition of a new compound that, it is hoped, will deal with the specter of possible drug resistance. But Marvin is grateful for every day of life, and happy to have his story told as a strong message of hope to all those who believe that they have been given a hopeless diagnosis.

Marvin's story brings into sharp focus the conclusion that cancer classification in the future may rest not upon what organ was involved, or what the cells looked like under the microscope, or where they had spread—but rather on a detailed molecular characterization of what cancer genes were involved.

PERSONALIZING CANCER TREATMENT

We have come a long way in our understanding of cancer. We now know that cancer is a disease of the genome, and that it comes about because of mutations in specific genes that promote inappropriate cell growth. The specific genes involved in certain tumor types are being cataloged at a furious pace. We are using that information to develop new ideas about targeted therapy, and we are learning that every tumor is a little different from the others.

Already, that information is being used to carry out molecular reclassification of tumors. Sometimes we can connect tumors on the basis of shared molecular abnormalities that we used to think had no relationship whatsoever. In an increasing number of cases, the information allows more accurate prognosis than any prior tools, and may even assist in the design of a specific therapy for that individual instead of the "one size fits all" approach.

It is not hard to see where this is going. In the not too distant future, every cancer will be characterized in great detail at the molecular level, soon to include complete DNA sequencing. The catalog of mutations for each of them will be spelled out. By cross-matching the pathways that are affected in a cancer with the inventory of therapeutics aimed at those pathways, a personalized and highly effective approach to treatment can be derived.

The future, thus, will be likely to include combination therapy of designer drugs, since cancers in general will have multiple abnormal pathways. The most successful approach will probably be to target as many of those as possible.

This era of personalized cancer therapy is already arriving. As one final dramatic story of a life saved by this new approach, meet Kate Robbins.

In the late summer of 2002, Kate Robbins began a "death journal." She had been diagnosed with metastatic cancer, and it was showing signs of rapid progression despite two major operations, radiation, and chemotherapy, so she decided to record daily events, to give her 9-year-old daughter and 11-year-old son something to remember her by. She wrote about her daughter's love of horses and her son's enthusiasm for Little League baseball. Writing was painful but cathartic.

Kate had been diagnosed with cancer on Valentine's Day. Kate, a nurse, and her husband, Mark, a radiologist, became concerned when she developed a headache that would not go away, and an MRI scan revealed a brain tumor. As part of her pre-op analysis before undergoing brain surgery, she was found to have a very large mass in her right lung. Ironically, it was her radiologist husband who first saw this scan and broke the news to her. Clearly, the brain tumor was actually a metastasis from the lung. Just 44 years old, Kate was a vegetarian who had never used drugs or tobacco, and so this discovery of metastatic lung cancer seemed utterly inconceivable.

Not being one to accept a hopeless situation, and determined to tackle this challenge in the most optimistic way, Kate underwent brain surgery, chest surgery, multiple rounds of aggressive chemotherapy, and radiation. Despite all that, she was found to have new metastases in the pancreas and liver by late summer. That's when she began keeping her death journal. Yet she never considered backing off, and when the possibility of a clinical trial for a new drug called Iressa was proposed, she jumped at the chance. Even if this doesn't help me, she thought, perhaps it will help someone else.

She started taking a daily Iressa pill at the beginning of 2003. She had lost 40 pounds, felt quite weak, and was actually a bit disappointed because the pill looked like a vitamin. Could it possibly provide the strong medicine she was hoping for? Yet a month later, her CT scans showed that her tumors had stabilized. She began to feel a little stronger. Amazingly, by May some of her tumors had actually disappeared. She and her physician, Dr. Tom Lynch, of Massachusetts General Hospital, marveled at this dramatic response, especially since other patients on this drug had not all had this same kind of benefit. An answer gradually emerged: this particular drug was designed to block the action of EGFR, a protein that is activated in some cancers. When the gene for this protein was sequenced in Kate's lung, it turned out to have a very specific mutation that made it particularly susceptible to the actions of the drug. Similar mutations were found in roughly 10 percent of patients in the trial, and they were the ones who had a strongly positive response to Iressa. The other 90 percent did not harbor such mutations, and generally failed to respond. So here was a dramatic example of personalized medicine. In this case, Kate had won the DNA lottery.

This was new territory, and the durability of Kate's response was completely unpredictable. But six years after her original diagnosis Kate has scans carried out every two months, and there has been no evidence of tumor in her lungs, liver, or pancreas. Her one remaining vexing problem is residual small metastases in the brain, which have escaped the effects of the drug because of the blood-brain barrier. This is a normal feature of human biology, and prevents many drugs from reaching adequate levels in the brain.

Nonetheless, the brain metastases have been progressing very slowly, and have been managed with surgery when needed. When I spoke with Kate, she was articulate about her story, grateful for the time that she has been given, and joyful that her son has now entered college and her daughter is not far behind. She stopped writing her death journal five years ago.

There is an ironic twist to this story regarding our medical system for approval of drugs. The FDA originally approved Iressa for the treatment of lung cancer back in 2003, on the basis of reports of dramatic responses such as Kate's. But the FDA insisted that a large randomized trial be conducted, comparing Iressa with the standard management of lung cancer.

In that trial with hundreds of individuals, Iressa showed no advantage over standard therapy when all the patients were considered together, and so the FDA withdrew its approval in 2005. Fortunately, the FDA agreed that this drug could continue to be made available to individuals like Kate who had already experienced a beneficial result. It also continues to be available in other countries. In a recent trial in Japan for patients with proven mutations in the *EGFR* gene, 63 percent of patients responded quite positively. Clearly, there is a lesson here: as personalized management of cancer becomes more and more practical, the FDA system for reviewing whether or not to approve a drug needs to take specific genetic data into consideration. A drug that may fail for 90 percent of cancer patients could still be lifesaving for 10 percent. If that group can be identified prospectively by DNA analysis, this is a drug that should be rapidly approved and made available to that subset.

The importance of considering individual differences in making decisions about cost effectiveness of medical interventions has taken on new importance as rising health care costs in the United States have led to a call for "comparative effectiveness research." Proponents of this approach argue that we can no longer afford the cost of all possible medical interventions for all individuals, regardless of whether or not evidence exists for their utility. Instead, they argue, research studies should be conducted to compare the available interventions in a given medical situation. Based on the results of such studies, only the option with the best overall outcome would be routinely reimbursed in the future. But if such comparative studies fail to take account of individual differences, as was the case above, in which the FDA withdrew approval for Iressa, then considerable harm will have been done to the promise of personalized medicine. Comparative effectiveness research makes scientific sense only if DNA studies are included in the study design, in order to identify a subset of individuals that may have a dramatically beneficial outcome from a particular medical intervention that otherwise seems inferior.

When she was first diagnosed in 2002, Kate undertook a search to try to identify someone who managed to achieve long-term survival after being found affected with metastatic lung cancer. She was dismayed to find not a single individual who had lived more than two years. If we consider the advent of targeted cancer therapy, her own persistence in seeking out all possible options, and some good fortune, Kate has now become the person she was searching for.

CONCLUSION

Of the many diseases that raise fears in ourselves and our families, cancer tops the list. Coming like a thief in the night, this culprit regularly steals away hopes for a long and happy life, afflicting its victims with loss of strength, loss of appetite, wrenching pain, and premature death. It is no secret that the methods used to treat cancer, including radiation and chemotherapy, have very serious side effects, since they not only attack the rapidly growing cancer cells but also affect normal cells in the body.

But the effort to catch and convict the culprits is rapidly gaining ground. The ability to search the genome for both hereditary and acquired mutations provides us with an increasingly precise picture of how these "genes gone bad" carry out their dastardly deeds. And learning their MO provides us with the opportunity to thwart their attacks in much more effective ways, including efforts to prevent the crime rather than trying to clean up the mess afterward. As the individual cases we have heard about in this chapter demonstrate, "law and order" is now a real possibility.

So the human genome is turning out to represent a powerful and personal textbook of medicine. But that's not all. It's also a personal history book. Written into your DNA are the life stories of your ancestors. What you can learn from reading those stories may explode your assumptions about who you are and how you are related to your fellow human beings. Are you ready for that kind of extreme makeover?

WHAT YOU CAN DO NOW TO JOIN THE PERSONALIZED MEDICINE REVOLUTION

1. Here's some advice that predates the genomic revolution but bears repeating everywhere and always. If you are a smoker, by far the most important step you can take to reduce your risk of cancer, heart disease, and emphysema is to quit. Smoking is addictive, and quitting is hard. But help is available! Start at http://www.cancer.gov/cancertopics/smoking and follow the many helpful ideas there (including free personal counseling from the National Cancer Institute) that have helped millions kick the habit.

2. Many women are deeply concerned about their risk of breast cancer, since this disease now affects one in eight women at some point in their lifetime. Genetic risk factors like *BRCA1/2* are providing opportunities to identify those at highest risk. To assess your own risk on the basis of family history, age, history of breast abnormalities, age at first menstrual period, and first delivery (if any), go to http://www.cancer.gov/cancertopics/factsheet/estimating-breast-cancer-risk. If your risk is substantially higher than one in eight, talk with your doctor. For more information about the *BRCA1/2* test, see http://www.cancer.gov/cancertopics/factsheet/Risk/BRCA.

3. Colorectal cancer is another condition for which early diagnosis provides the best opportunity for cure. Currently it is recommended that all individuals undergo regular colonoscopy starting at

age 50. The National Cancer Institute has recently posted an online tool that allows individuals to assess their risks, based on family history, diet, amount of exercise, and use of tobacco. See http://www.cancer.gov/colorectalcancerrisk/.

4. If you have a family history of cancer, or any indications of heightened personal risk based on previous genetic testing or early warning signs, you should consult with your physician to be sure you are taking advantage of all possible methods for surveillance and early detection. You can stay current with the latest developments in cancer prevention and genetics at http://www.cancer.gov/cancertopics/prevention-genetics-causes.

CHAPTER FIVE

What's Race Got to Do with It?

At age 51, Wayne Joseph was a successful black professional who had worked hard to make a life for himself and his family. He went to a black high school, married a black woman, and raised both his daughter (named Kenya) and son to be proud of their black heritage. Working through the public high school hierarchy to become a principal, he was a pillar of the African-American community. But at the same time, he looked forward to a day where race would not be such a focus of attention, publishing a somewhat controversial article in *Newsweek*, "Why I Dread Black History Month," in which he argued that it would be more useful for black people to promote a color-blind society than separate out their accomplishments.

When Wayne saw an advertisement from a company called AncestryByDNA, he was intrigued by the claim that it would help him discover his original heritage in sub-Saharan Africa. And he thought it might make a good topic for another essay. So he signed up.

When the results came back, Wayne was shaken to his foundation. His DNA analysis report indicated that he was 57 percent Indo-European, 39 percent Native American, 4 percent East Asian, and ZERO percent African.

He had heard that his mother's family included people who "crossed the color line," but zero percent African? He checked to be sure the company had not mixed up any samples—it had not.

He then concluded that he must be adopted—but when he confronted his mother, she quickly disabused him of that idea. He found his birth certificate; in the box for race it said "Negro." And his coffee-colored skin certainly fit with that assumption. But as he learned more about his family's history in New Orleans, it became clear that there were many interesting and diverse roots to his family tree—and perhaps none from Africa after all.

The reaction of others to this news was disbelief. "They wanted to keep me in the same box I had always occupied," he told me. His mother said, "I'm still a colored woman, I'm too old and too tired to change." Wayne's brother said that the test hadn't been done on *his* DNA, so therefore he was still black. Wayne's son and daughter were shocked. Some of his black friends teased him about his essay in *Newsweek* from a few years earlier: "We should have known you weren't really black," they said. High school students seeking to disregard his instructions on their proper conduct were heard saying, "He never seemed black to us." Most dramatically, Wayne's second wife, who was white, said, "You have to be black; I defied my mother to marry you!"

Over the five years since this news reached him, Wayne has gradually come to peace with his identity—but he says it helped that he was already 51 years old when he learned this information, and that he "would have needed therapy" if he had made the discovery 30 years earlier. Not long after his discovery, his mother, his wife, and his best friend all died within a period of nine months. Wayne noted that none of them talked about their race when they were dying; it just didn't matter very much.

Wayne has attempted to use his own experience with DNA testing in conversations with students in the black high school to argue that race is irrelevant—but identification with race is important to them, and they are not ready to give it up.

WHAT DOES RACE MEAN, ANYWAY?

What's going on here? Is it really possible for a DNA sample to provide this kind of precise window into ancestry? Can this result be trusted? And what does our DNA tell us about our traditional ways of categorizing ourselves?

Almost all of us are rapidly acculturated as children to notice physical differences between groups, and before long we learn various social labels that are applied to ourselves or to others and are based on skin color, hair texture, facial features, language, ancestry, or culture. Yet, as nicely expressed by Evelyn Brooks Higginbotham, "When we talk about the concept of race, most people believe that they know it when they see it, but arrive at nothing short of confusion when pressed to define it."

Perhaps it would be good to try this experiment right now yourself. If pressed to write down a definition of *race*, as it applies to humans, what would you say?

The word "race" does have a scientific definition of sorts: "a geographically isolated breeding population that shares certain characteristics in higher frequencies than other populations of that species." Hardly a transparent definition! But even aside from the question of which characteristics and populations should qualify, there's a problem—with very few exceptions, humans don't hang out for long periods in geographically isolated groups. After all, if you started walking in eastern China and progressed steadily westward to the tip of Portugal, you would not encounter any precise dividing lines between individuals with different physical characteristics; nor would you encounter any absolute barriers against interbreeding. Despite this, biologists who were classifying species of plants and animals attempted to define racial classifications for humans in the 1600s. Linnaeus developed four human racial classifications: Americanus, Europeaus, Asiaticus, and Africanus. The characteristics that Linnaeus assigned to each of these races reflect the strong prejudices of his time and place; Europeaus were said to be "acute, inventive, gentle, governed by laws," whereas Africanus were described as "crafty, indolent, negligent, governed by caprice or the will of their masters."

It got worse. Johann Blumenbach, building on Linnaeus's classification, coined the word "Caucasian" to describe white-skinned individuals, providing a source of endless confusion from that point onward. After all, there has never been any evidence that individuals of this skin tone derived from the Caucasus Mountains; Blumenbach simply chose that term because he thought individuals in that geographic area were particularly attractive physically. Most destructive of all, Blumenbach took the horizontal arrangement of races that Linnaeus had proposed, and arranged them vertically, with Caucasians at the top.

There is absolutely no scientific justification for such an arrangement. All humans alive today have been derived by way of branches of equal length from our common original ancestors. But Blumenbach's view, which found its way insidiously into European thinking, contaminated human beings' understanding of one another over the course of the next three centuries, and still reverberates today. In the words of Stephen Jay Gould, "The shift from a geographic to a hierarchical ordering of human diversity must stand as one of the most fateful transitions in the history of Western science—for what, short of railroads and nuclear bombs, has had more practical impact, in this case almost entirely negative, upon our collective lives?"

Is there any biological basis whatsoever for current racial categories? The short answer is no, since our racial categories are vague and heterodox. "African-American" is a label that obscures much more than it explains—and many dark-skinned people, like Wayne Joseph, have no recent connection to Africa. Our DNA *can* tell us something about geographic ancestry, however. Some variations in human DNA are a historical record of our ancestors' migrations around the planet. These are now being utilized in the offering of DNA testing by commercial enterprises, including the one that gave Wayne Joseph such a surprising result.

In our society, especially in the United States, the concept of race is freighted with considerably more baggage than just biological ancestry. Racial designations such as "African-American," "Alaska Native," or "Asian" include critically important nonbiological components such as history, language, and culture. In this chapter, we will unpack some of those issues, and then come back to how this may or may not be relevant to personalized medicine.

ALL HUMAN BEINGS HAVE PROFOUNDLY SIMILAR DNA SEQUENCES

The DNA sequence of any two individuals is 99.6 percent identical, regardless of which parts of the world their ancestors came from. Compared with the rest of the animal kingdom, humans are unusually similar: most other species have considerably more diversity in their DNA sequences than this. Most of the genetic variations in our species are found to be present within geographic populations. Only about 10 percent of those differences are useful in predicting which populations you belong to.

I recall the first time that I shared with President Clinton this information about humans' similarities. During his presidency, and since then, Clinton has been deeply interested in the Human Genome Project. (He says learning about it helps put him in touch with his "inner nerd.") Realizing immediately the social significance of these profound DNA similarities, Clinton lectured to the Serbs and Croats in Kosovo about the intrinsic irrationality of their ethnic war! I am not sure how much of an impact his arguments had upon them, but he was making a good point. If we humans choose to hold on to racial prejudices, or even to make war on each other, like the Serbs and Croats, or the Protestants and Catholics in Northern Ireland, or the Tutsi and Hutu in Rwanda, we cannot use biology as a justification.

This profound similarity between individuals at the DNA level is a reflection of the relatively recent arrival of humans on Earth. Population geneticists, looking at the variation in the human genome across the world, have concluded that all 6 billion current members of *Homo sapiens* are descended from a common set of approximately 10,000 ancestors. These founders of the human race lived 100,000 to 150,000 years ago, most likely in East Africa. Much of the variation that can be measured today in humans was already present in those 10,000 founders, but has been reshuffled and recombined along the way as groups migrated from one place to another.

Other branches of the genus *Homo*, such as the Neanderthals, apparently split off from our lineage about 500,000 years ago, and ultimately became extinct. Recent success in determining the DNA sequence of the Neanderthal genome (from ancient bones) so far shows no evidence of interbreeding with *Homo sapiens*, even though both species apparently coexisted in Europe about 30,000 years ago.

THE FREQUENCY OF CERTAIN GENETIC VARIATIONS CAN DIFFER ACROSS THE WORLD

Even though most common genetic variations that we find today existed in our common ancestors, our DNA diversity has not been evenly distributed since humans spread from eastern Africa. Figure 5.1 is a graphical depiction, based on DNA evidence, of how human migration must have occurred. Migrations of groups of humans out of Africa into Europe and Asia may have involved relatively small numbers of ancestors. When one of those ancestors happened to carry a variation that was relatively uncommon in Africa, it would then become much more common in the new land. This so-called *founder effect* can account for significant differences in frequencies of certain genetic variants, and could provide an ancestral signature for future offspring.

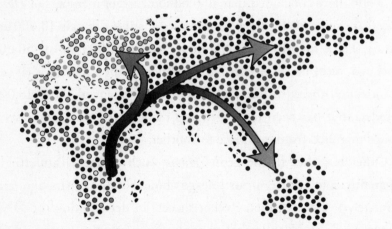

Figure 5.1: Schematic of the flow of human genetic variation out of Africa over the last 30,000 years. The greatest degree of variation still exists in Africa, with subsets of that having been carried by groups that populated Europe and Asia.

In addition to these founder effects, a phenomenon known as *genetic drift* also resulted in random slow changes in the frequency of certain variations over time. Thus a variant that might have been present in 20 percent of East Africans could end up being present in 40 percent of Asians, purely on the basis of random chance. Although one such variant would tell you very little about an individual's likely ancestry, a large collection of such variants with skewed distributions could provide a robust statistical estimation of that person's ancestry. To the extent that new mutations were also cropping up in individuals who had already left Africa, those could also provide information about population history.

SELECTION HAS SHAPED SOME PARTS OF THE HUMAN GENOME

Careful consideration of events over the last 50,000 years makes it clear that evolution has continued to leave its mark, even in more recent times. In some instances the force of natural selection has been supplied by the physical environment. Perhaps the most obvious is skin color. There is good evidence that the original ancestors of *Homo sapiens* were dark-skinned. Such skin pigmentation was, in fact, a necessity for hairless individuals living near the equator and roaming out in the savanna. Dark skin was critical for protection from ultraviolet radiation. Without that, skin cancers would have developed early in life and would have been devastating in consequence. As a volunteer physician in Nigeria in the 1990s, I cared for several patients with albinism who had suffered that consequence. They appeared at the hospital with horribly advanced skin cancers as early as age 20, because of their lack of necessary skin pigmentation.

On the other hand, individuals who migrated out of Africa to more northern latitudes would have run into trouble maintaining adequate levels of vitamin D. That process requires absorption of sunlight through the skin. Thus dark-skinned individuals are prone to vitamin D deficiency in areas of limited sun exposure, causing a disease called rickets. In its full-blown form, rickets results in distortion of the skeleton, and a high likelihood of difficulties during pregnancy and delivery that can be fatal to both mother and baby. Thus, individuals with lighter skin would have had more reproductive success in northern climates, and ultimately light skin became predominant.

Recently, a major molecular cause of this change in skin color has been identified for Europeans. Specifically, the gene *SLC24A5* turns out to be critical for the production of melanin, the predominant dark pigment of the skin and hair. In Africans, as in most other vertebrates, that gene is fully functional. But virtually 100 percent of Europeans have a mutation in *SLC24A5* that greatly impairs the function of the protein. Thus it is fair to say that all white-skinned Europeans like me are mutants! Interestingly, Asians have the fully functional version of *SLC24A5*, but have acquired mutations in other genes that result in lighter skin, while retaining black hair.

Another force of natural selection has been the human diet. The most dramatic example relates to the ability to digest milk sugar (lactose) in adulthood. The evidence suggests that in our common African ancestors, the ability to digest milk was turned off at about age two. At that point, the *enzyme* called lactase stopped being produced in the intestine in sufficient amounts to allow lactose to be digested. In the absence of lactase, a significant amount of milk in the adult diet results in bloating, diarrhea, and inability to gain nutrition from the milk sugar. But certain groups in Europe, the Middle East, and East Africa gradually developed agricultural practices that included the keeping of cows and goats, from which they obtained milk. Those who could digest the lactose in the milk had an advantage. So now in those populations one finds that most adults continue to produce lactase. The molecular basis for this developmental change has been identified. Most northern Europeans carry a specific mutation in a regulatory signal for the lactase gene that allows it to stay "on" throughout adulthood. Certain milk-drinking African tribes such as the Maasai also show persistent lactase expression, but they turn out to have a different mutational reason for it. This is an excellent example of how natural selection has operated in recent human history. Finding two different molecular changes to achieve the same outcome in two different populations is referred to as "convergent evolution."

Another strong force of natural selection that continues to operate on the human genome is the need to resist infection. We have already mentioned the prominent distribution of the sickle-cell mutation in individuals from West Africa, driven by the protection against malaria that the sickle trait provides. In fact that same mutation is also found around the Mediterranean, mapping rather precisely onto the worldwide distribution of the malarial parasite over the last several thousand years. In other malarial areas, different "helpful" mutations have also reached high frequencies, again driven by the protection that carriers receive against early death from infection by this aggressive blood parasite. Other evidences of selection operating upon the genome in West Africa can be pinpointed to genes that have provided resistance to Lassa fever. Some people have speculated that the worldwide epidemic of HIV/AIDS will eventually leave its mark also, as those genetically predisposed to resist infection may have a higher likelihood of reproductive success.

These are all very small differences, among the 10 percent of our human variations that can be ascribed to our geographic roots. What about more controversial differences? Is there any variation in, say, athletic prowess or intelligence based on geographical pressure? Well, it's clear that differences in physical stature can contribute to athletic abilities—descendants of the Maasai, with their traditionally tall stature, are more likely to succeed in basketball than their neighbors the African Pygmies. As for intelligence, however, there is currently *no* evidence that genetic contributors to intelligence are differentially distributed across the globe.

DNA CAN MAKE PREDICTIONS ABOUT ANCESTRY

If you handed me four DNA samples, and said one came from a person who lived in Japan, another from Spain, another from Nigeria, and a fourth from a Native American living in Arizona, I could go to the laboratory, spend a little time doing DNA analysis, and almost certainly tell you which was which. But my success would depend upon the fact that each of those individuals had ancestors who had lived in those geographic areas for some time, so that their DNA reflected the features of those founders.

If, on the other hand, you gave me a DNA sample from the golfer Tiger Woods, I would have a more difficult time. By his own description, Woods is one-quarter Chinese, one-quarter Thai, one-quarter African-American, one-eighth Native American, and one-eighth Dutch. Nonetheless, by testing a sufficient number of DNA variations that are known to have somewhat different frequencies across the world, I could probably make a reasonable guess about his mixed ancestry. Wayne Joseph, whose story began this chapter, had his own mixed ancestry assessed by just this kind of DNA analysis. And though the percentages in his report could be off a bit, the basic conclusion of his analysis is probably about right.

In some instances, however, the commercial business of testing for ancestry has gotten a little ahead of the science. Some testing services even claim to be able to tell African-American individuals from which African village their original slave ancestors came. That could be correct only if relatively little migration had occurred within Africa itself during the past few thousand years. Such precise geographic conclusions would also require very comprehensive DNA sampling across all the villages of Africa, which is not yet available.

As the ability of DNA analysis to predict ancestry has grown in accuracy, this approach has begun to find its way into forensics in new and controversial ways. Recently, law enforcement officials in Louisiana were on the trail of a serial killer, from whom they had derived a DNA sample from a bit of material left at the scene of the crime. Eyewitnesses disagreed about the physical features of the suspect, some reporting that he was black whereas others said he was white. Using a psychological profile developed by the FBI, the authorities focused on searching for a white male aged 25 to 35. But a DNA diagnostics company called DNA Print was called in. It analyzed the sample and said the perpetrator was 85 percent sub-Saharan African and 15 percent Native American, and would be expected to be dark-skinned. The police work then shifted to a different list of suspects. Eventually a black male was apprehended, and his DNA was found to match that collected at the crime scene. After a court trial, he was convicted of murder, and he is now serving a life sentence.

Some would say that this was a valuable adjunct to police work, since it led to an arrest and a conviction. However, given our inability to make precise predictions, one could also imagine an alternative scenario in which such information could throw law enforcement off the track and lead to harassment of innocent parties.

This kind of "DNA profiling" is likely to become more and more prevalent as time goes on. Scientists are now in the process of identifying DNA variations that play a role in facial features, hair texture, and adult height. Might it be possible in the future that the police graphics artist will depend as much on the DNA sample as on the description of eyewitnesses?

THE INCONGRUITIES OF RACIAL CLASSIFICATIONS

Given the long history of gene flow among various groups that make up *Homo sapiens*, the notion that one could precisely define a subset of individuals and say that they are somehow separate from the rest of the human race is clearly not scientifically defensible. The history of the human species over the last 100,000 years is sometimes depicted as a branching tree, but this image implies that the branches are separated from one another. We are much more of a trellis than a tree; or perhaps a wisteria vine is a better metaphor. Nonetheless, there is a long tradition in many societies, often driven by specific social prejudices, that attempt to divide up *Homo sapiens* into a series of non-overlapping subgroups. Perhaps nowhere have the consequences of this "racializing" of society been more apparent or more hurtful than in the United States, where the great sin of slavery continues to reverberate despite almost 150 years of emancipation.

The prejudicial effects of slavery on racial categorizations are immediately apparent in the "one-drop rule," which was articulated early in U.S. history. Many children born as the result of exploitation by white plantation owners of slave women were assigned the racial status of black slaves. Though these children were half European, this assignment provided economic advantages to the plantation owners, as well as sustaining the racist philosophy of white superiority. The one drop rule was carried to an extreme, requiring that any individual with even a single black ancestor many generations back be designated as black. Now that we know we are all descended from black African ancestors, the basis of the one-drop rule can be seen as utterly absurd. Yet it was used for decades as a means of maintaining the economic and social superiority of one group over another.

The United States government has its own confused history in dealing with race. Over the decades, the U.S. Census has adopted many different approaches to try to place people in racial categories. Currently, the Office of Management and Budget (OMB) lists five different races for census purposes: American Indian/Alaskan Native; Asian; Black or African-American; Native Hawaiian/Pacific Islander; or White. In addition, individuals are categorized by whether they are Hispanic/Latino or not, so that an ethnic label is put on top of the five racial labels. Individuals are allowed to designate their own race and ethnicity, and to mark more than one category. A tip of Uncle Sam's hat is now given to the realization that these categories have no scientific basis; OMB states that "these should not be interpreted as being primarily biological or genetic in reference."

The arbitrariness of these classifications becomes immediately clear when one looks around the world. As a consequence of the U.S. history of slavery and the one-drop rule, individuals like President Obama, who is 50 percent African and 50 percent white European in ancestry, are referred to as black or African-American. In Brazil, however, only individuals of predominant African ancestry and very dark skin are considered black, so Obama would be considered white. Surely this points out how completely meaningless these terms are!

HEALTH DISPARITIES

So should we get rid of racial designations once and for all, given that their biological basis is so shaky, and that they carry a history of discrimination? Such a decision would no doubt help diminish prejudice. However, there is one significant medical argument why this should not be done, at least not now: health disparities.

A health disparity exists when one population group experiences a higher incidence of, or higher morbidity or mortality from, a specific disease, compared with the general population. There are many troubling and long-standing examples. Consider just a few. Prostate cancer is considerably more common and more lethal in African-American males than in European or Asian males. Type 2 diabetes, which is becoming increasingly common in all groups, is particularly prevalent and serious among Native Americans and African-Americans; more than 50 percent of Pima Indians in the southwestern United States are diagnosed with type 2 diabetes by the age of 50. Stomach cancer occurs with significantly higher frequency in Asians than in Europeans or Africans. Crohn's disease (inflammatory bowel disease) is seen more commonly in Europeans than in other groups. If we believe that one of our goals as members of the human species is to improve human health for all, ignoring these health disparities would be unforgivable.

As a result, like it or not, we will probably need to continue to consider various methods of population categorization for some time, at least until we have sorted out the reasons for these differences in health outcomes and implemented measures to correct them.

One should not assume, however, that health disparities are caused by genetic differences. As pointed out already, genetic differences between groups are going to be small, and many other factors that have nothing to do with DNA often play a much more significant role. Such environmental factors as socioeconomic status, educational opportunity, exposure to toxic environments, access to health care, cultural practices, diet, and even stress due to discrimination could lead to disease. Sorting out these multiple factors contributing to health disparities is currently a high-priority goal for biomedical research.

The diagram in Figure 5.2 shows the possible ways in which self-identified race or ethnicity may connect to health outcomes. Both environmental and genetic factors may be involved, in different proportions for each disease.

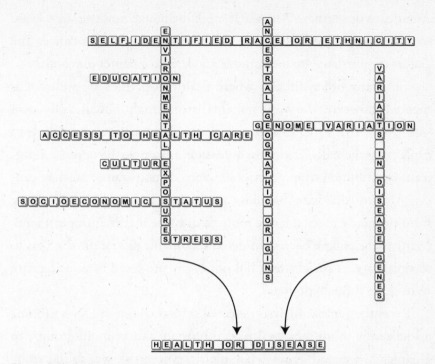

Figure 5.2: A graphic depiction of the complex connections between self-identified race and health.

Still, although non-genetic factors no doubt account for many health disparities, in a few instances one can point to genetics as a major factor. That has already been discussed for a number of rare recessive diseases such as cystic fibrosis (which affects primarily Europeans), sickle-cell anemia (which affects primarily Africans), or Tay-Sachs disease (which affects primarily Ashkenazi Jews). The high prevalence of those conditions in particular geographic areas can be explained by founder effects or natural selection. In at least one instance of a common disease, the health disparity can be immediately explained on the basis of biological differences: dark skin is protective against all forms of skin cancer, including the highly malignant cancer called melanoma, and so it is clear why the incidence of those cancers is much higher in individuals of European background than in African-

Americans or southern Asians. (It might be noted, however, that when dark-skinned individuals in the United States do get skin cancer, the cancer is more likely to metastasize and lead to a poorer outcome.)

For many other diseases where health disparities are well documented, however, the basis remains frustratingly obscure. My own suspicion is that genetics will turn out to play a relatively small role in explaining these disparities. In at least one example, however, a significant contribution from genetics has been demonstrated: prostate cancer. African-American men have the highest rate in the world, dying from prostate cancer at a rate more than twice that of European men. Many theories have been considered that blame diet, limited access to medical care, or social stress, but none have produced convincing data to back up those hypotheses.

Recently, genetic analysis detected several common DNA variants associated with an increased risk of prostate cancer in all groups. In Chapter 3, we read about Jeff Gulcher, whose risk was found to be elevated by testing for these variants.

A cluster of these risk variants lies in a stretch of about 1 million DNA base pairs on chromosome 8. That cluster contains no fewer than seven independent risk variants, each of which can raise the risk of prostate cancer by anywhere from 10 percent to 30 percent. To the surprise of investigators, the risk variant in all seven instances is substantially higher in frequency in African-American men than in European or Asian men. Remarkably, it appears that this one region on chromosome 8 may account for a substantial fraction of the health disparity in prostate cancer between Europeans and African-Americans. If confirmed, this would be the first example in which a health disparity for a common disease appears to be strongly driven by genetics, though it is also likely that environmental factors are contributing in ways that have not yet been fully defined.

CAN RACE BE A PROXY FOR PERSONALIZED MEDICINE?

A few years ago, a practicing physician created an uproar by writing an op-ed in the *New York Times* titled "I Am a Racially Profiling Doctor." The author, Dr. Sally Satel, explained how she used the apparent racial background of her patients to make decisions about which drug to give for heart disease, depression, hepatitis infection, or pain relief. In each of those instances, she cited published data to show that as a group African-Americans and Europeans had slightly different responses to those therapeutic interventions. Thus she argued that her kind of medical racial profiling was evidence-based and motivated by a desire to practice the best medicine for the benefit of all of her patients.

There is a certain thread of logic here. After all, no one would have objected if Dr. Satel had indicated that she is much more observant about the possibility of malignant melanoma in her white patients than in her darkest-skinned African-American patients. The problem with her presentation about the other applications of race-based medicine, however, is that these conclusions are based upon relatively small differences between groups, and may therefore be utterly irrelevant to the individual.

Let's look again at Figure 5.2. Although the existence of health disparities demonstrates a connection between the top of the diagram (self-identified race and ethnicity) and the bottom of the diagram (health outcome), a real desire to help the individual requires focusing on the steps in between, in order to individualize the treatment more effectively. Thus while race may be a proxy for those more significant and proximate factors, it is a lousy proxy, and will often give inadequate information that may actually limit the ability to practice good medicine.

Unfortunately, the intermediate factors, both environmental and genetic, have still not been identified for many health disparities. So for the time being, there will still be instances where we have no better options, and Dr. Satel's recommendations will need to be considered. However, the medical care system should collectively make the commitment to moving as quickly as possible to get beyond the imperfect and potentially prejudicial proxy of race, and to dissect out the individual causative factors that are really influencing health. Important among those is the family medical history, which is too often ignored, but generally provides much more specific guidance than any generalizations from race or ethnicity.

RACE AND RX

A particularly instructive example of the complexities and misunderstandings of race-based medicine is the story of BiDil, a drug for the treatment of congestive heart failure. This was the first drug specifically approved by the FDA exclusively for African-Americans, and it caused both celebration and consternation.

BiDil is actually a combination of two generic drugs that have been available for decades at very low cost. This combination, a single pill, is sold under the patented brand name BiDil at a substantial cost markup. The drugs work to dilate blood vessels, reducing the amount of resistance the failing heart has to pump against. The scientific rationale for the treatment of heart failure with this "afterload reduction" approach is quite strong, and clinical studies began as early as in the 1970s. In the 1980s, two trials of this drug combination were conducted on several hundred male veterans, and showed modest evidence for overall decreased mortality. But the difference between the treated and untreated groups was not quite statistically convincing, and the FDA denied approval of the drug combination. The company that was pursuing approval abandoned the project.

Ten years later, another company (NitroMed) picked up the drug combination. Its interest was based upon a new analysis of the data from the original trials, which indicated that African-American veterans might have actually had a significant benefit from the drug, even though it did not measure up in the combined study of blacks and whites. This new information was used as the justification to file for a new patent on this drug combination, now called BiDil, since the old patent was approaching expiration. The U.S. Patent Office issued the patent, the first race-specific patent ever given for a drug. NitroMed then asked permission to market the drug specifically for the treatment of heart failure in blacks. The FDA responded that such approval could be granted only if a large new clinical trial was conducted solely on African-Americans, to see whether BiDil really worked.

More than 1,000 African-Americans with congestive heart failure were recruited, representing a substantially larger group than the original trials from the 1980s. The results were so encouraging that the trial was stopped early: black men and women receiving BiDil experienced a 43 percent reduction in deaths over two years. On the basis of these results, the company quickly completed its request for approval of BiDil for the treatment of congestive heart failure in African-Americans. The FDA agreed, and inserted this language in the drug label: "BiDil is indicated for the treatment of heart failure as an adjunct to standard therapy in self-identified black patients."

This was unprecedented. After all, most clinical trials of drugs are conducted primarily on white people, and the FDA never says on the label of an approved drug that it won't work for other groups.

The FDA decision split the community asunder. Many blacks, including black professionals, hailed this development as the first time that a therapeutic for a serious medical condition had been developed for them. "Finally!" they said. But others in the black community decried this race-specific outcome as scientifically unfounded, creating the unfortunate and inaccurate impression that self-identifying as African-American makes one biologically different from some other group.

In retrospect, the original trials in a mixed population were probably not large enough to discern whether or not some benefit might have also accrued to nonblack patients being given this drug. The only large definitive trial was conducted solely on African-Americans. Thus it was not really legitimate to say that this drug would not have worked for other groups as well. On top of that, there were aspects of this story that caused concern about whether this was more about profit than about medical benefit. After all, issuance of a race-based patent permitted extension of patent coverage for many years, preventing the development of a generic drug that would have saved patients a great deal of money. To make the profit motive and the heavy marketing of the drug to physicians treating African-American patients even more troubling, BiDil was a fixed combination of two inexpensive generic drugs. So exactly the same therapy can be given at greatly reduced cost, if a physician is willing to write two prescriptions, and patients are willing to take a few more pills.

Sales of BiDil have failed to meet NitroMed's expectations, perhaps because of these confounding facts. But the impression that it would be possible to develop race-specific medicines has lingered on. In many people's view, including my own, the BiDil story has left an unfortunate legacy.

CONCLUSION

There are no human races in the strict biological sense. We humans represent a wonderful continuum of marvelous diversity, all descended over just a few thousand generations from a common pool of black African ancestors. Truly, we are all one family. But each of our DNA instruction books carries the footprints of our history over the last 100,000 years, and careful examination of our specific sets of genetic variations can reveal that. Furthermore, since hereditary factors play a role in almost all diseases, and genetic variation is not uniformly distributed across the world, at least some part of our future risks of illness are probably in some way connected to the location where our ancestors lived over the last few thousand years. Thus an effort to understand health disparities requires a close examination of DNA. None of that justifies the definition of subgroups of individuals as biologically distinct races. Furthermore, "race" as used in common parlance carries many other connotations that go well beyond biology to include culture, history, and social status. Though we might ultimately hope to see race eliminated as a consideration in human society, an insistence on that outcome right now would cause us to ignore health disparities, and to offend many individuals for whom self-identified race is an important part of their personal identity.

The goal for personalized medicine must be to move as swiftly as possible toward the identification of individual risk factors, be they environmental or genetic, that play a direct role in disease risk. Racial profiling in medicine, even if well intentioned right now, should recede into the past as a murky, inaccurate, and potentially prejudicial surrogate for the real thing.

WHAT YOU CAN DO NOW TO JOIN THE PERSONALIZED MEDICINE REVOLUTION

1. In 2008, a group of 18 geneticists, social scientists, lawyers, and ethicists published the "Ten Commandments of Race and Genetics." Have a look at those at http://www.newscientist.com/article/dn14345-ten-commandments-of-race-and-genetics-issued.html.

2. Explore the science of ancestry testing more deeply. National Geographic and IBM are sponsoring the "Genographic Project," which is sampling thousands of individuals around the globe to build a DNA history of our species. Check out http://video.nationalgeographic.com/video/player/specials/in-the-field-specials/grand-central-genographic.html.

3. For a thoughtful statement about the science of ancestry testing from the American Society of Human Genetics, see http://ashg.org/pdf/ASHGAncestryTestingStatement_FINAL.pdf.

CHAPTER SIX

Genes and Germs

At 42, Uri Davis (not his real name) was much too young to be facing two fatal diseases. But here he was, an American living in Berlin, already dealing with AIDS and now diagnosed with a rapidly progressive form of leukemia. He was like many others with AIDS since the 1990s: his infection had been kept pretty much under control by triple-drug therapy. But his new and very serious problem—leukemia—had failed to respond to chemotherapy. Uri's options were running out.

His physician, Dr. Gero Hutter, recognizing his patient's unusual plight, hatched a plan that might prove to be a double whammy for Uri's pair of diseases. But the stakes were very high. If chemotherapy fails, the recommended treatment for leukemia is a stem cell transplant, despite its high risk of complications and even death. But Dr. Hutter had another idea. He initiated a careful study of the 80 compatible stem cell donors, seeking to identify one who might be nearly ideal for Uri. Donor 61 had just the right characteristics, and so was chosen for the transplant.

At the time of the transplant, all of Uri's AIDS drugs had to be stopped, because they might damage the donated stem cells in the fragile period when the transplant was taking hold. The expectation was that the drugs would have to be restarted shortly afterward, or the AIDS virus would return with a vengeance.

The drugs were never restarted. Two years after the transplant, no evidence of HIV can be found in Uri's body. Although it is still possible that pockets of virus might be lurking somewhere, a scientific meeting convened to discuss his case concluded he is "functionally cured."

What happened here? Why are researchers calling Uri's case a "proof of principle" for a completely new therapeutic approach to AIDS? And what does all of this have to do with genetics and personalized medicine? To answer these questions, we need to delve into an intricate component of human biology: the immune system.

YOUR GENES AS DEFENDERS AGAINST PATHOGENS

Though you may not like to think about it, you are surrounded at all times by a sea of microorganisms. While many of them are benign, some are actually quite virulent, and almost all of them if introduced into the wrong place in your body would be capable of creating a serious infection. Yet for the most part you coexist quite comfortably with these organisms, with hundreds of trillions of them living in or on your skin, your mouth, and your gastrointestinal tract.

Fortunately, evolution has designed your body to be very effective at restricting the opportunity of these organisms to cause disease. For the most part your body's bloodstream is completely sterile. (There are some exceptions—for example, after a vigorous dental procedure, a transient burst of bacteria may be found in your blood, but it rapidly clears.) Important defenses against infection are provided by the mechanical barriers in the skin, the nose and mouth, the GI tract, and the vagina. You also possess a complicated and magnificent array of immune responses, mediated by various immune cells and proteins. They allow you to fight off those frequent episodes in which a microorganism has ventured into a space where it should not be.

Because all our biological functions are encoded by the genome, it should not come as a surprise that a significant fraction of your 20,000 genes are involved in the immune response. And since genetic variations occur in virtually all the genes in the body, it should also not come as a surprise that our immune systems are not all quite identical at birth, and that some of these variations can play a role in susceptibility or resistance to specific illnesses.

HIV/AIDS AND GENOMIC MEDICINE

In the middle of the twentieth century, there was great optimism that the scourge of infectious diseases might at last be coming to an end. The development of effective antibiotics against bacterial infections, as well as the introduction of vaccines against dangerous viruses such as polio, portended a future where viral and bacterial pathogens might no longer be a serious problem for humans.

That optimistic view came crashing down soon afterward. Resistance to antibiotics quickly emerged as a major problem, with microorganisms continually threatening to get ahead of the development of new drugs. And most devastating of all, a strange wasting disease of the immune system appeared in 1981 and gradually spread around the world as a full-fledged epidemic. That disease, which has now claimed the lives of more than 25 million people, is acquired immunodeficiency syndrome (AIDS), which is caused by the human immunodeficiency virus (HIV).

AIDS was first identified in the United States among young gay males in California. These men were afflicted with a variety of highly unusual infections and/or an unusual form of cancer called Kaposi's sarcoma, all indicative of an immune system that was utterly failing. Subsequently, the cause was traced to a virus. The disease was shown to be spread by sexual contact (both homosexual and heterosexual), through blood products or needle exchange, and from mother to child at the time of birth.

The virus proved to be diabolically clever. Its genetic information is encoded within RNA, not DNA. When the virus enters a cell, the RNA is converted into DNA, using an enzyme that the virus brings along to its own party. The viral DNA is then inserted into the genome of the infected cell, where it proceeds to make copies of itself. The predominant human cell type that is infected, called a T cell, is a major part of the immune system. So as the viral infection continues, the immune system is gradually destroyed. To make matters even worse, the virus has the capacity to mutate quite rapidly, so that the body's natural immune system is repeatedly thwarted because the virus "changes its clothes" to the point where it cannot be recognized.

Extensive epidemiological studies indicate that ancestors of the AIDS virus have been present in chimpanzees for a long time. Most likely the transfer to humans occurred sometime between 1884 and 1924 in Africa, perhaps during the butchering of chimpanzees for human consumption.

Without treatment, the median time from HIV infection until death from AIDS is approximately 10 years. In the 1980s and early 1990s there were few exceptions to this ultimately fatal outcome. In response to this rapidly spreading epidemic, an unprecedented public-private partnership was developed to seek some means of effective therapy for individuals infected with HIV. The result of this vigorous research program was the development of "highly active antiretroviral therapy" (HAART). This approach, which includes a combination of drugs in order to avoid the rapid development of resistance to any single drug, has provided a dramatic reduction in death rate. AIDS is now a chronic disease, not a certain death sentence, in countries with access to therapy. But these drugs do not represent a cure, and individuals who have stopped therapy have generally relapsed rapidly. Efforts to develop effective HIV vaccines have been extremely frustrating, despite a great deal of research. The rapid mutability of the virus and the body's seeming inability to mount an adequate immune response against it are major reasons for this lack of success.

Given these sobering circumstances, by far the most appropriate strategy for reducing the spread of HIV/AIDS is prevention. That has led to worldwide efforts to educate individuals about safe sex practices, especially the use of condoms to reduce transmission of the virus. These efforts have been only partially successful, however. With the success of HAART, there has even been a disturbing trend toward resurgence of risky sex practices in the developed world.

As the HIV/AIDS epidemic has continued to spread, researchers have been surprised by occasional individuals who were repeatedly exposed to the virus, yet never developed illness. Some of these were gay males with large numbers of sex partners. But the most dramatic examples were individuals with hemophilia, who had been exposed to hundreds of units of HIV-infected blood products at a time when screening for the virus was not yet possible.

The story of HIV and hemophilia is particularly tragic. Hemophilia causes repeated episodes of bleeding into joints and internal organs. Its inheritance is X-linked, so males are primarily affected. It can be effectively managed by infusing specific blood components from donors. Those infusions supply the blood-clotting factor that is genetically missing in these affected males. But preparation of concentrates of the needed blood products requires the use of large pools of donor units, so a single donor infected with HIV can transmit the virus to many recipients. This outcome was devastating for the hemophilia community in the early 1980s, before screening of the blood supply was available. A large fraction of hemophiliacs became infected with HIV and died.

But a few hemophiliacs who had been repeatedly exposed to HIV showed no evidence of infection. Researchers postulated that some genetic form of resistance might be present in these persons, and that it might provide important clues to how to prevent the disease in others. The hunt was on.

At the same time, other research groups studying how HIV infects specific immune cells figured out that the virus uses a docking procedure to get inside the cell. The virus can gain access only by binding to a group of proteins that lie on the surface of these particular immune cells (Figure 6.1). One of these proteins is called CCR5. Like all proteins, it is encoded by a human gene. When the researchers began investigating the *CCR5* gene in hemophiliacs who were AIDS-resistant, they were amazed to discover that many of them harbored a serious mutation in this gene: 32 base pairs of the gene were deleted.

Since proteins are made up of *amino acids*, and three base pairs of DNA code for any given amino acid, you could say, using language as a metaphor, that the "word length" of DNA is three. Three letters make each word; a string of words make each protein-sentence. Sentences can be 6, 9, 12, 15 (etc.) letters in length, comprising 2, 3, 4, 5 (etc.) words. This *deletion* of 32 base pairs is therefore a particularly severe one, since 32 is not divisible by three. We call this a *frameshift mutation* because it throws off the reading frame. If your DNA has such a mutation, iti slik ereadin gword slik ethes e. In fact, later work showed that this so-called *CCR5 Δ32* mutation results in a complete loss of the CCR5 protein.

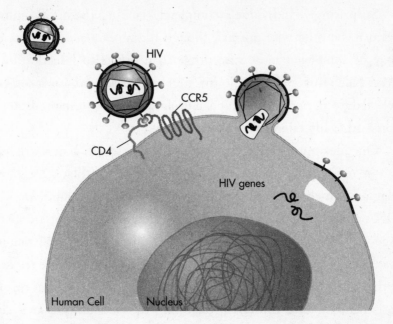

Figure 6.1: The diabolically clever AIDS virus (HIV) docks to the normal immune cell surface proteins CD4 and CCR5, and then gains access to the cell. It goes on to make many copies of itself, destroying the cell in the process. For an eye-catching animated simulation of these steps, see www.boehringer-ingelheim.com/hiv/art/art_videos .htm.

Many of the individuals with HIV resistance have actually inherited a copy of this deleted gene from each of their parents, and have no normal copies of the gene at all. Only about 1 percent of Europeans (and few if any Africans or Asians) have this complete loss of the protein CCR5, but these individuals are essentially protected from infection with most strains of HIV. Subsequent studies have indicated there are a few other genes that also confer relative protection, but none are as strong in their effect as this one. Approximately 12 to 16 percent of European individuals have one copy of the deleted gene as well as one normal copy, and they too seem to have some partial benefit, delaying the onset of full-blown AIDS by two or three years after HIV infection.

Surprisingly, individuals completely lacking the CCR5 protein seem to be otherwise normal, though recent evidence suggests they may be somewhat more susceptible than others to infection by the West Nile virus. But how did this mutation arise, and was it somehow selected for in northern European evolution by providing resistance to some other infection in the past?

One idea put forward was that in the fourteenth century this mutation might have provided protection from the Black Death (the plague), which killed 30 to 60 percent of Europe's population. But DNA analysis of bones from individuals who died during the Bronze Age (2,900 years ago) in Germany indicate that the Δ32 mutation was already present, and at about the same frequency as we see today. Other ideas have been put forward about the advantage that this mutation might have conferred, including the possibility that it could have provided resistance to smallpox. But those hypotheses have not been proved.

So now let's return to the story of Uri Davis. Knowing about CCR5 and the Δ32 mutation, Dr. Hutter sought to identify a stem cell donor with two copies of this mutation. He hypothesized that if these stem cells were successfully transferred to Uri, the donor immune cells would be able to resist the further spread of HIV. And it worked—as the stem cell transplant took hold, the evidence of HIV in Uri's body disappeared. Dr. Hutter was amazed. This was the most dramatic demonstration that blocking CCR5 in an individual already infected with HIV could score a knockout of the virus, perhaps indefinitely.

There has been considerable scientific discussion about Uri's case. No less an authority on HIV/AIDS than the Nobel laureate David Baltimore has pronounced this a "proof of principle" for gene therapy as a possible approach to treating AIDS. Here's the idea: stem cells or bone marrow would be removed from an infected individual, treated with a recombinant DNA vector to turn off the *CCR5* gene, and then returned to the patient. These cells should then have a selective survival advantage over the untreated cells. This is still a hypothetical approach, and may well run into technical difficulties, but it is one of the more promising therapeutic ideas of the last few years.

Several pharmaceutical companies have also taken advantage of these observations to pursue development of drugs that bind to the CCR5 cell surface protein in normal individuals, preventing the virus from docking to the cell. One of these drugs is called Selzendry (its generic name is maraviroc). It was approved by the FDA for the treatment of AIDS in 2007 and shows considerable promise.

The HIV story discloses many features of the interactions between a pathogen and the host, and how genetics plays a role. And close inspection of other infectious diseases reveals many similar themes.

MALARIA

When I volunteered as a missionary physician in Nigeria 20 years ago, I saw firsthand the ravages that this parasitic disease can cause. The diabolical cycle involving the malaria parasite, the mosquito, and humans represents the greatest current infectious scourge of humankind. A person living in a malaria-infested area is bitten by mosquitoes continually, especially at night, and will experience infection hundreds of times during a lifetime. The human toll is extremely high.

During one of my visits, both my daughter and I were infected. We experienced the profoundly debilitating effects of this disease, thankfully truncated by the availability of effective drug therapy. Unfortunately, though, all too often therapy is not available for those of limited means in the countries where the disease is rampant. Children are particularly susceptible to a lethal outcome from infection by the malarial parasite, because they have not had the opportunity to develop the partial immunity that arises from repeated infections. Nearly 1 million people, most of them children, will die this year of malaria in sub-Saharan Africa.

As with HIV, there are clearly genetic factors that play a role in the likelihood of severe infection with the malarial parasite. We have already noted that the sickle-cell mutation has risen to a high frequency because of its ability to confer relative resistance to malaria for those who carry one copy. But that is not the only such mutation that has been selected in malarial regions. Another blood condition, beta-thalassemia, which is particularly common around the Mediterranean, also involves production of hemoglobin in the red blood cells, where the malarial parasite likes to hang out. Carriers of beta-thalassemia also seem to benefit from resistance to malaria. In Southeast Asia, a related condition, alpha-thalassemia, occurs at high frequency and presumably also confers relative resistance to malaria.

In addition, individuals who lack a particular enzyme, called G6PD, in their red blood cells, or who have low levels of this enzyme, do better after exposure to malaria. Because the gene for the enzyme resides on the X chromosome, males are most commonly found to have this mutation. They seem to suffer few ill effects unless they encounter certain dietary substances, such as fava beans, for which the enzyme is needed to avoid a nasty case of anemia. A long list of drugs can also create serious problems for males with this mutation.

Yet there is much hope for the development of new methods of prevention and treatment for malaria. In addition to the use of public health strategies, such as bed nets, to prevent mosquito bites at night, molecular biology and genetics are coming to the fore in the design of new approaches. Now that the complete genome sequence of many different malarial strains has been determined, it is possible to begin to assemble better information about which parts of the parasite are invariant and, therefore, most vulnerable to vaccine preparation. The availability of the DNA sequence of the malarial genome is also spurring the development of new antimalarial drugs. Real progress is being made. Very recently, a malaria vaccine is showing promise by reducing the incidence of infection, portending a possible future in which this disease may actually be eradicated.

TUBERCULOSIS (TB)

As a medical intern in Chapel Hill, North Carolina, in 1977, I was urgently called to the outpatient clinic one day. A young migrant worker had collapsed in the bathroom, surrounded by a prodigious amount of bright red blood. Moving him quickly to the intensive care unit while trying to bring his blood pressure back up with fluids, and obtaining blood for transfusion, I was mystified by why a young individual would have had such a massive intestinal hemorrhage. As he began to regain consciousness in the ICU, my patient's deep rattling cough prompted me to grab for a mask. But it was too late. A few months later, a skin test showed that I had contracted tuberculosis.

This case ended up causing quite a stir at our medical grand rounds. My patient's dramatic gastrointestinal hemorrhage was caused by tuberculosis that had spread throughout his body, resulting in an erosion from his splenic artery directly into his colon. He also turned out to have extensive cavitary TB of the lung, and readily passed a few of those "red snappers" (as we interns referred to the TB bacteria) to me.

Many people consider tuberculosis a disease of the past. It certainly was a terrible scourge in previous centuries, taking the lives of young and old, and sending many others to sanatoriums, where the intentions were good but the means for treating the disease were limited. The introduction of antibiotics in the 1940s made a substantial difference in the otherwise gloomy picture, but because the TB organism grows so slowly and is so effectively walled off in little pockets in the lung, treatment often has to be carried out for many months, making management difficult. (I took a drug called INH for a full year.)

Tuberculosis also likes to infect compromised hosts. With the increasing incidence of HIV/AIDS, TB is making a frightening comeback. The concern is heightened all the more by the increasing appearance of organisms that are resistant to virtually all known anti-TB drugs.

A great deal of effort is currently going into developing new antibiotics and vaccines for TB. Here again, the genomic strategy, which involves obtaining the complete DNA sequence of a large number of different TB strains, may provide the best hope of identifying the greatest vulnerability of the organism. Meanwhile, investigators are searching for genetically encoded host factors that may play a role in susceptibility. Recently, this search has been rewarded. A gene called *SLC11A1* has been shown to harbor a common genetic variation that plays a role in susceptibility. Another gene, called *TLR2,* seems to assist in containing the infection, as individuals with a less functional variant of *TLR2* are predisposed to the most lethal form of tuberculosis: TB meningitis. As with HIV/AIDS, the discovery of these host factors has provided important clues to the development of new strategies for prevention and treatment.

INFLUENZA

The worst of all recent global pandemics was the "Spanish" influenza epidemic of 1918–1919. During these two years an estimated one-third of the world's population was infected. Between 50 million and 100 million people died. Other influenza epidemics had occurred before this time, and more have happened in waves ever since, but the emergence of this particularly virulent strain was unprecedented.

Like HIV, and like most viruses, influenza is capable of gradual mutations that change its biological character. Presumably in 1918 mutations arose that conferred particular virulence and transmissibility from the normal host (birds) to humans. But genetic factors also played a role in who lived and who died. Researchers in Utah, studying genealogical records and death certificates, have been able to show that the relatives of someone who died of influenza were themselves more likely to die, even after accounting for their shared environment. The exact nature of the genetic predisposition to the severe outcome has not yet been defined. It may ultimately be possible to define this, however, since in Utah genealogies are carefully maintained, and the DNA sequence of deceased individuals can be reconstructed by studying their relatives.

The world has been holding its breath for the past four or five years, fearing that another influenza pandemic might be just around the corner. A strain of influenza denoted H5N1 is decimating bird populations, particularly chickens in Southeast Asia. This virus is also occasionally capable of causing human disease for those who have had heavy exposure to infected birds. Some 250 fatal cases have been reported. There are great fears that as little as one or two more mutations in the avian viral genome might unleash a frightening scenario of person-to-person spread. But so far there are only 36 examples of an originally infected person who has spread the disease to other family members. Interestingly, in only four of those clusters is an unrelated person (a spouse) involved. In all of the other instances, the infected individuals are related, suggesting that they may share some gene variant that makes them particularly vulnerable to H5N1. Discovering the nature of this variant is a high priority, as it may provide an important clue to prevention and treatment, should this disease emerge again in pandemic proportions.

Just to prove that nothing about influenza is entirely predictable, the rapid emergence of a new strain of the virus in Mexico in March 2009 caught everyone by surprise. This particular virus, classified as an H1N1, seems to be an unusually mongrelized mix of genomic segments from viruses previously circulating in birds, pigs, and humans—so it's probably unfair to our porcine friends to call it "swine flu." The virus has rapidly spread to many countries, proving its high human-to-human transmissibility. So far, however, the mortality rate from this virus does not appear to be exceptionally high, providing some reassurance that a replay of 1918 is not at hand. There can be no assurance, however, that the virus will not acquire greater lethality over the coming months, so a major effort to prepare and administer a specific vaccine is under way.

THE HUMAN MICROBIOME PROJECT

As a final consideration for this chapter, let me introduce you to a new area of scientific research that is likely to have profound consequences for personalized medicine in the future. In this case, the studies are not about the human genome, but about the genomes of the microbes that live on and in us. These microbes actually outnumber us. Your body is made up of approximately 400 trillion cells. But if you add up the number of microbial cells on your skin, in your mouth and nose, and in your intestinal tract, the total comes up to 1 quadrillion (1,000 trillion, or 1,000,000,000,000,000). Not only are there more microbial cells than human cells, but these microbes are enormously diverse. And so the total number of genes they carry is wildly in excess of our modest list of just 20,000 genes.

It is appropriate, therefore, to begin to think of human beings as superorganisms, existing in a symbiotic relationship with these microbes. For the most part they contribute to our normal health and well-being, because they have been adapted to us, and us to them, over the course of millions of years.

But this symbiosis can be disrupted. Most researchers suspect that such disruptions may cause more examples of illness than we are currently aware of. A major problem is that many microbes cannot be isolated and studied in the laboratory, as they seem able to grow happily only in their human host. Thus, we have only a sketchy view of how disruptions of the normal microbiome may actually be causing a long list of diseases.

A recent, stunning example is stomach ulcer. For decades, such ulcers were considered to be a matter of too much gastric acid, often blamed in turn on stress. Treatment, often unsuccessful, was based on bland diets and antacids. More recently, it has become clear that stomach ulcers are actually caused by the bacterium *Helicobacter pylori*. Thus the appropriate treatment of ulcers is not antacids but antibiotics!

I strongly suspect that a number of other mysterious diseases may turn out to be caused in large part by disruptions of the microbiome. Likely to be on the list are chronic gum disease (gingivitis), inflammatory bowel disease (Crohn's disease and ulcerative colitis), and various other intestinal, skin, and vaginal infections.

With the advent of high-throughput, low-cost *DNA sequencing*, a new window is opening for the investigation of the role of the microbiome in health and disease. Even though many of these organisms may fail to grow in the laboratory, they have DNA, and so their presence can be deduced by carrying out extensive DNA sequencing of material derived from various body sites. Already, interesting results are appearing in cases of a childhood skin problem, eczema, which may have more to do with change in the skin microbiome than has been previously realized.

A particularly surprising finding is that the microbial constituents of the intestinal tract may play a fundamental role in obesity. Recent studies of lean and obese individuals indicate a substantial difference in the microbes represented there. In experiments with mice, transferring the microbes from an obese mouse to a lean mouse results in weight gain for the recipient, suggesting that the microbes are in some synergistic way participating in the efficiency of utilization of calories. These results have electrified biomedical researchers, demonstrating that new strategies for dealing with obesity might be developed by modifying intestinal flora.

THE FUTURE

With regard to the topics in this chapter—in contrast to previous chapters—there are currently relatively few specific genetic tests that an individual can take to warn of future risks of infectious disease. One possible exception is the availability of testing for the *CCR5 Δ32* mutation, to determine susceptibility to AIDS. But here the value of knowing one's status is of questionable significance. In fact, alarm bells sounded when advertising for testing for this mutation began appearing in gay magazines. The concern was that those found to have two copies of *CCR5 Δ32* might then relax their concern about their risk of AIDS, and indulge in risky sex. That would obviously be an unfortunate choice, since *CCR5 Δ32* does not provide protection for a long list of other sexually transmitted diseases.

It is safe to predict, however, that in the future this kind of information about genetic susceptibility to infectious disease will play a role in personalized medicine. This will come about in several ways:

1. Within the next few years, complete genome sequencing for many of us will include the discovery of various factors that will play a role in predicting susceptibility to illness. This might be valuable if, for instance, you are traveling to a malarial area and want to get a sense of how serious an exposure would be for you.

2. Prediction of vaccine responses will also become possible. Not everyone given the proper dose of a vaccine against a particular pathogen has the same response. Genetic factors play a large role in that. In the future, this will provide an opportunity to optimize the dose or the frequency of administration, on the basis of the individual's genetic makeup.

3. It is highly likely that sampling of one's personal microbiome will become an important part of the diagnostic workup for certain diseases such as skin rashes, vaginal infections, and intestinal upsets. More than that, it is not hard to imagine a time in the future when routine sampling of the microbiome from various body sites will be utilized as an early warning sign of trouble, even before symptoms have appeared.

4. If you do acquire an infection in the future, it is likely that the drugs prescribed will be based on genomic insights about your genes and the genes of the pathogen, as is already happening in the example of *CCR5* and HIV.

5. Treatment of infectious disease with drugs will continue to be the mainstay for most infections. However, response to these drugs may vary from individual to individual. Increasingly, genetic testing will play a major role here, paving the way for the choice of the right drug at the right dose for the right person. I will have much more to say about this in Chapter 9.

This exploration of infectious disease has carried us a long way from the classic version of genetics Mendel would have recognized.

Hold on to your hat, because we're headed into even woollier territory. Consider this question: might significant features of your personality, like whether you prefer adventure or have a tendency toward depression, also be written within your genes?

WHAT YOU CAN DO NOW TO JOIN THE PERSONALIZED MEDICINE REVOLUTION

Personalizing your approach to avoiding infectious disease includes knowing and practicing the principles of safe sex. There is currently no cure for HIV, and many other sexually transmitted diseases can be very serious and difficult to treat, so your best strategy is to do everything possible to avoid becoming infected. A trusted place to obtain valid medical information on this topic and many others is MedlinePlus, operated by the National Library of Medicine. See http://www.nlm.nih.gov/MEDLINE PLUS/ency/article/001949.htm for its recommendations on safe sex.

CHAPTER SEVEN

Genes and the Brain

You have probably seen stories in the media about sensational findings linking specific genes to various brain diseases, and even to personality traits or behaviors. But it's easy to slip into a false sense that we will find a simple genetic explanation for everything we do. So here's a quiz: Which of these things are almost entirely determined by genes? Alcoholism, Huntington's disease, spirituality, depression, sexual orientation, marital fidelity, intelligence. Hint: Only one. Now, to which of these do genetic factors make a contribution? Hint: This is a book about heredity!

I started medical school just as recombinant DNA was being invented. The possibility of being able to reshuffle bits of DNA was electrifying the scientific community and worrying the public. With a new PhD in chemistry, I was drawn to the beauty and elegance of the DNA molecule, and resolved within a few months of entering medical school that genetics would be my focus.

Medical genetics was still an arcane discipline, however. I sought out the one individual in my medical school who was interested in the genetics of common diseases, and followed him around as much as I could. One afternoon, I asked him whether there were any adult-onset conditions in which heredity was absolutely irrelevant. He responded that there were likely to be quite a few, but the most compelling example was Parkinson's disease, which had been clearly demonstrated by numerous research studies to have absolutely no genetic contributions. With its late onset and generally sporadic occurrence, Parkinson's disease must be caused entirely by some unidentified environmental exposures. Or so he said.

Fast-forward 20 years, to 1994. As the director of the National Human Genome Research Institute at the National Institutes of Health, I was approached by researchers studying Parkinson's disease. Frustrated by their inability to identify its causes, they wondered whether a genetic approach might potentially be useful.

Recalling my former mentor's comments, I thought this sounded like a dead end. But my friend and colleague Dr. Robert Nussbaum took a strong interest in the problem. Within a year, I was standing in Bob's laboratory as he showed me the data pinpointing the precise genetic cause of Parkinson's disease in a group of Italian and Greek families. These families included many individuals who were stricken quite early, in their forties or fifties. Their symptoms were otherwise typical: tremor, rigidity, a loss of facial expression, and difficulty initiating movement, leading to falls and ultimately to using a wheelchair or being bedridden. The inheritance pattern in these families suggested that a single dominant gene mutation must be present.

That's what Bob's team had identified. There in the middle of the coding region of a gene called alpha-synuclein, a single letter was misspelled. The result was a benign-appearing change in a single amino acid. But the change was sufficient to cause this devastating disease in those who carried the mutation.

Shortly after that, it became clear that the alpha-synuclein protein is a major contributor to the degeneration of neurons in a particular part of the brain, the substantia nigra, which is invariably affected in Parkinson's disease. Neurons in this part of the brain produce a substance called dopamine as their neurotransmitter. A mainstay of treatment for Parkinson's disease is the replacement of the missing substance by the drug L-dopa.

But actual mutations in alpha-synuclein are rare causes of Parkinson's disease. You can, therefore, think of alpha-synuclein mutations in Parkinson's disease as analogous to *BRCA1* mutations in breast cancer, causing only a small percentage of the cases. What about the rest? Subsequent work on other families has revealed no fewer than 13 separate places in the genome where subtle changes are capable of increasing the risk of Parkinson's disease. One of the most common is the *LRRK2* mutation found in Google founder Sergey Brin and his mother (see Chapter 3).

Was my genetics professor at medical school completely off base? Is Parkinson's disease actually completely genetic, and is environment unimportant after all? Not so fast. The very year I was having this discussion with my professor, something shocking and astounding was happening in another academic institution a few hundred miles away.

Barry Kidston was a 23-year-old graduate student in chemistry at the University of Maryland in 1976. In his high school senior photo, still retrievable on the Internet, Barry has the look of a budding young scientist of the 1970s, complete with uncombed hair and horn-rimmed glasses. But his interest in chemistry was apparently more than just academic. Barry set out to synthesize a compound called MPPP, an analog of the narcotic Demerol, intended for his personal consumption. It was supposed to give him a high akin to heroin. He injected himself with the results. But apparently something had gone terribly wrong, as three days later Barry developed full-blown symptoms of Parkinson's disease. His unusual case led to an investigation by the National Institute for Mental Health (NIMH), which discovered that there were traces of another related compound, MPTP, in the glassware that he had used for the synthesis. The sleuths at NIMH suspected that MPTP might be toxic to the dopamine-producing cells in the human brain, but they were unable to show that it caused any damage to laboratory rats. Depressed by his permanent and serious neurological problems, Barry sank further into drug addiction and ultimately died of a cocaine overdose.

Six years later, it happened again: seven young addicts turned up in emergency rooms in California with acute Parkinson's disease. This time, sleuths tracked down the source of their particular illegal drug, made in a garage and known on the street as "China White." Once again, significant contamination with MPTP was discovered. This time the compound was tested in primates, and was found to induce severe symptoms of Parkinson's disease. (Another lesson: rats are useful proxies for humans in many respects, but not all.)

Poisoning with MPTP is a very rare cause of Parkinson's disease. But these two stories—the Italian and Greek families with alpha-synuclein mutations, and Barry's chemistry experiment gone wrong—define the boundaries of the full spectrum of Parkinson's disease, from a completely genetic cause to a completely environmental cause. Most cases lie in between. In this regard, Parkinson's disease can be considered as a model for nearly all common human illnesses. Early detection of risk, before irreversible organ damage has occurred, will be critical. As we move ever closer to a full understanding of personalized medicine, each individual case of illness will need to be properly placed along the spectrum of genetic and environmental causes in order to design the most effective treatment.

THE MOST COMPLEX ORGAN

Besides being Woody Allen's second favorite organ, the human brain is the most complex organ in any creature on Earth. It is estimated to contain 50 billion to 100 billion neurons, which pass signals to each other through about 100 trillion connections. That a mere 20,000 human genes, contained within a 3 billion-letter DNA instruction book, are sufficient to specify such a marvelous organ continues to amaze all those who think about it in more than a superficial way.

The brain expresses most of these 20,000 genes—hardly any are always "off" in the lifetime of the brain. The subtleties of their expression in different parts of the brain are only just now beginning to be worked out. Some of the incredible anatomic complexities of the brain are provided by an elegant dance of developmental timing, in which genes turn on or off at just the right moment. This complexity is enhanced by the fact that many of the genes in the brain do not specify just one protein but use "alternative splicing" to create more than one. One remarkable gene in the brain is estimated to be able to make 38,000 different proteins!

Brain development is not entirely hardwired by the genome. Of all the organs in the body, the brain is the one where environmental interactions are most crucial. External stimulation during early childhood is essential for normal brain development, producing a profound effect on which connections get made or broken. Even in adulthood, the experience of learning new information results in changes in neuronal connections. These connections can also be disrupted by illness or drugs.

Given the enormous complexity of the brain, the underlying tapestry of individual genetic variation, and the sweeping influence of environmental exposures, it is no wonder the brains of any two individuals—even identical twins—are profoundly different. Unfortunately, however, some of these differences can lead to serious illnesses.

NEURODEGENERATIVE DISEASES

My father had been a folk song collector in the 1930s, and during my childhood in the 1950s traditional folk musicians often stayed at our farmhouse in the Shenandoah Valley of Virginia. I learned early to love the songs of Woody Guthrie, America's foremost balladeer. But Woody's voice was prematurely silenced while I was still a child. He became progressively impaired with a mysterious disorder that caused erratic behavior, uncontrollable jerky motions of the extremities, and ultimately a slow downhill course to his death in 1967. The disorder was Huntington's disease, formerly called Huntington's chorea. It had taken his mother's life when Woody was only 15. Two of his daughters later developed it—though thus far his son Arlo, a famous singer in his own right, appears to be spared.

Huntington's disease (HD) is a classic example of a dominantly inherited adult-onset condition. Each child of an affected individual has a 50 percent chance of inheriting the genetic mutation and developing the disease (see Figure 2.3). The average age of onset is 37. In 1993, my laboratory was part of a collaborative effort that discovered the mutation that causes the disease. An intense effort has continued since then to develop better means of treatment. But currently, although there are promising leads, researchers are still struggling to achieve the much-hoped-for breakthrough.

Huntington's disease is just one entry in a long list of highly heritable neurodegenerative diseases. Many are rare. One of them, Charcot-Marie-Tooth disease, was referred to in the introduction, because it affects my father-in-law and may have been inherited by my wife. In addition, there are more complex neurological disorders whose inheritance cannot generally be so easily understood, though clearly genetic factors are involved. One of these is Parkinson's disease, which we have already discussed. Another is Alzheimer's disease, which we will explore further in the chapter on aging.

MAJOR MENTAL DISORDERS

The body's most complicated organ is also the site at which disruptions of function can cause a great deal of disturbed behavior and human misery. Worse yet, mental illnesses often have been stigmatized in the past, placing even more burdens on affected individuals and their families. Mental illnesses often appear in young people, and are best understood as developmental disorders rather than neurodegenerative conditions. They remain a challenging frontier of medical research, as the causes, diagnostic methods, and treatment strategies remain far from optimal.

We will consider four specific mental disorders, and then briefly summarize what is being learned, from studies of the human genome, about their possible causes.

Schizophrenia

In contrast to what you might think from novels and movies, schizophrenia is *not* a disorder in which the individual has a "split personality" (that's properly called "multiple personality disorder"). In reality, this common disorder (about one in 100 individuals is affected) is characterized by hallucinations, delusions, and disorganized and unusual thinking. A somewhat more accurate, but still highly romanticized, version of schizophrenia was depicted in the movie *A Beauti-*

ful Mind, chronicling the life of John Nash, who won the Nobel Prize in mathematics in 1994. Sadly, schizophrenia is generally much more devastating than it was for John Nash. Though medications that control some of the more troubling symptoms have been available for 50 years, and have allowed the release of many affected individuals from mental institutions, these do not resolve the cognitive problems, and so do not represent a cure. The swelling of the ranks of the homeless by schizophrenics is a reflection of our collective failure to come up with adequate solutions to this devastating disease.

Manic-Depressive Illness (Bipolar Disorder)

In her moving book *Touched with Fire*, Dr. Kay Jamison quotes the poet Lord Byron at the opening of her first chapter: "We of the craft are all crazy. Some were affected by gaiety, others by melancholy, but all are more or less touched."

In the same book, Dr. Jamison, herself affected with bipolar disorder, goes on to analyze the lives of major literary and artistic figures of the last several centuries. She concludes that a substantial proportion of them have had manic-depressive illness. Although creativity can reach remarkable peaks during the manic phase, the depths of depression that accompany this illness in its full-blown form represent the kind of blackness that can hardly be imagined by an unaffected person, and have all too frequently led to suicide. About 1 percent of individuals in all societies appear to be affected by this condition, and even though the introduction of lithium treatment has done much to limit the extremes of the emotional highs and lows, the condition is still poorly understood.

Major Depressive Disorder

Nearly everyone will experience occasional periods of situational depression after discouraging life events. But the onset of major depressive disorder may have no precipitating factors; the affected person will experience prolonged sadness, irritability, sleep disturbances, loss of appetite, and loss of the ability to experience pleasure, extending over more than two weeks, and interfering significantly with work and relationships. This is the most common of the major mental illnesses, affecting more than 12 percent of men and more than 20 percent of women during their lifetime. Major depressive disorder has a variable course and an unpredictable response to treatment, but nearly all patients with this distressing disorder can be helped once the diagnosis is recognized.

Autism

The tragedy of autism is profound. Currently about one in 150 children receives this diagnosis, with males affected about four times more frequently than females. Diagnoses of autism have increased significantly in the last 30 years, but there is considerable controversy about whether this trend represents a real increase in incidence, or simply greater awareness of autism among parents and health care professionals. Generally starting before the age of three, the condition is characterized by impairment of communication and social interactions, as well as restricted and repetitive behaviors. Given the age of onset and the apparent increase in incidence, much concern has been raised about the possibility that immunizations, especially those containing the preservative thimerosal, might play some role in the initiation of autism. However, repeated objective analyses by panels of distinguished experts have shown no evidence for this connection.

As just one piece of evidence, thimerosal was removed from all vaccines in the United Kingdom in 1991, and yet the incidence of autism in the United Kingdom has continued to rise, just as in the United States. The continued promotion of a connection between vaccines and autism, spurred on by a highly vocal faction of the autism advocacy community, is hard to fathom in light of these objective research studies. Though it is understandable that parents of children with autism are desperately seeking answers, this continued propagation of a discounted theory is now beginning to threaten the utilization of childhood immunizations, and new epidemics of measles are already appearing as a consequence.

Genetic Factors in the Major Mental Disorders

For all of these conditions—schizophrenia, bipolar disorder, major depression, and autism—there is strong evidence of genetic factors. Studies on that revealing experiment of nature, *identical twins*, show strong concordance for all four of these conditions. The identical twin of a schizophrenic has a 50 percent chance of also being diagnosed with the disease, whereas a *fraternal twin* has only a 15 percent chance. In bipolar disorder, the identical twin concordance is about 60 percent. In major depressive disorder, it is about 40 percent. In autism, depending on the study and the precise definition of the disease, the concordance is as high as 90 percent. Though one could argue that a shared intrauterine environment might explain these high levels of concordance without implicating heritable DNA per se, the fact that fraternal twins also spend nine months within the same uterus and yet have much lower rates of concordance suggests that heritability is crucial.

So far, despite the use of techniques for studying common variations by genome-wide association studies (GWAS), described in

Chapter 3, the mutations in these four diseases remain mostly in the "dark matter" of the genome. There have been a few confirmed findings: in schizophrenia, a variant in a gene called *ZNF804A* appears to increase risk slightly; in bipolar disorder, genes known as *ANK3* and *CACNA1C* have been implicated; and in autism the *SHANK3* and *CAD10* genes have shown evidence of involvement. But these findings do not account for more than a small percentage of the risk.

Perhaps we should have been prepared for this outcome, given evolutionary considerations. Recall that the GWAS approach is capable of detecting only *common* variants in the human population. A genetic variant that is associated with a significant reduction in reproductive fitness will never rise to the level of being common in the population. It is clear that autism and schizophrenia substantially reduce the likelihood of achieving parenthood. Given this circumstance, one might predict that the genetic factors in autism and schizophrenia might instead be relatively rare, arising often as new mutations, resulting in a relatively high risk of mental disorder, but disappearing from the population after a few generations.

Some evidence is accumulating that this model may be correct. Specifically, those large-scale rearrangements of the genome known as copy number variants (CNVs; see Figure 3.7), in which segments of repeating code within the genome are altered by adding or deleting copies, have been identified in schizophrenia and autism at a significantly higher frequency than in normal individuals. In some instances, these CNVs are large enough to have resulted in the duplication or deletion of entire genes, adding to the likelihood that they might have some significant effect on function. Given that the brain is the most complicated organ, it is perhaps not surprising that this kind of genome reorganization may have its most serious consequences in the nervous system.

What is needed to resolve this mystery is the complete DNA sequence of individuals with these disorders. That would allow discovery of newly arising CNVs and other rare variants that have a large effect. Until this work is completed, it will be difficult to say with any confidence whether these disorders represent a vast collection of different diseases, or whether a common molecular theme will emerge, leading to better understanding, prevention, and treatment.

Meanwhile, those interested in genetic diagnosis or testing for susceptibility to mental illness should be aware that a few such tests are already being marketed, but their validity remains questionable. For example, a company called Psynomics is marketing a DNA test for susceptibility to bipolar disorder, arguing that this information could be useful in establishing the diagnosis in an uncertain case. The test being offered, however, is based upon a variation in a gene called *GRK3*, and this has not been validated in a large-scale study. The result could turn out to be utterly useless. Even worse, this kind of unvalidated test, utilized by individuals or their physicians to make a serious diagnosis in an uncertain situation, might do more harm than good. This is one more argument why additional oversight of genetic testing is needed.

BEHAVIORAL GENETICS

What about personality traits and behaviors that aren't so rare or debilitating? Even here, genetic factors turn out to be important. Furthermore, some of these traits may intersect with the medical care system; they may influence our response to personalized medical recommendations; they may be included in DNA test results from companies that are marketing directly to consumers; and they are certainly conditions about which people of all ages are curious. None of what follows is intended to medicalize personality, or diminish free will. If anything, understanding the genetics of personality should *empower* free will.

Depression after Major Life Events

Major clinical depression was discussed above. But what about episodes of situational depression? Individuals apparently differ in their ability to recover from events such as breakups of romances, deaths, illnesses, or job crises. Do genes play any role in these responses?

Increasingly, we are learning of specific examples in which that is the case. A study of twentysomethings in New Zealand revealed that 17 percent of them had experienced an episode of depression in the preceding year. The investigators speculated that individual differences in serotonin, one of the feel-good chemicals in the brain, might be involved. Interestingly, when they studied variations in the gene for a serotonin transporter protein in the brain, they found a significantly higher likelihood of a depressive response in those who had two copies of the so-called short allele, a variant that reduces serotonin uptake. Even more provocatively, they showed that individuals with the short allele who had also experienced child abuse were particularly susceptible to these episodes of depression.

If this conclusion is correct (and that is not certain, as other recent studies have failed to replicate this result), this seems to be an example in which the genes we inherit provide a foundation upon which our environment operates, in a fashion that Matt Ridley details in his fascinating book *Nature via Nurture*. Ridley writes compellingly about the interactions between genes and environment: "Genes are not puppet masters nor blueprints. Nor are they just the carriers of heredity. They are active during life; they switch each other on and off; they respond to the environment. They may direct the construction of the body and brain in the womb, but then they set about dismantling and rebuilding what they have made almost at once—in response to experience. They are both cause and consequence of our actions. Somehow the adherents of the 'nurture 'side of the argument have scared themselves silly at the power and inevitability of genes, and missed the greatest lesson of all: the genes are on their side."

Though Ridley wrote these words prior to the discovery of the role that genes play in the response to stressful life events, he might very well have been writing about this very phenomenon.

Alcohol Dependence

Alcoholism causes untold misery in our society—from deaths of innocent people at the hands of drunk drivers, to families torn apart by alcohol abuse, to the enormous health care expenditures arising from the consequences of alcoholism. It is assumed by many that alcoholism is purely and simply a matter of choice. Certainly that is a critical factor in determining whether an individual falls prey to this addiction. After all, the incidence of alcoholism in the Old Order Amish community is extremely low. Yet it is also clear that individual susceptibility to alcoholism varies on a genetic basis. Studies of identical twins raised in separate environments confirm the fact that there must be genetic factors involved. Two such genes have been identified, and provide a useful model.

Figure 7.1 shows the metabolic pathway by which ingested alcohol is converted to acetate, which can then be used by the body as an energy source. This conversion travels through an intermediate step, the formation of acetaldehyde. Acetaldehyde is related to formaldehyde, the substance traditionally used for preserving cadavers in the anatomy lab. Acetaldehyde is toxic, creating very unpleasant sensations if it rises to significant levels, including skin flushing and nausea. In effect, a person with high levels of acetaldehyde is being "pickled."

Figure 7.1: Enzymatic steps involved in the metabolism of alcohol.

Alcohol itself is capable of inducing a pleasant experience of "being high," but to the extent that acetaldehyde builds up, the experience is overshadowed by less pleasant symptoms. Ultimately, when all the alcohol has been metabolized to acetate, the experience is over. Think of alcohol ingestion in the same way as riding a roller coaster at an amusement park: the exhilaration of speed, sudden drops from great heights, and a titillating sense of danger create a "high" for many people. But for some of us, the initial high is replaced by a growing sense of discomfort and nausea, and when the ride is over, we are only too happy to get off and find a quiet place to recover.

Once alcohol is consumed and the "high" begins, one enzyme catalyzes the first stage of digestion, creating acetaldehyde, and creating the poisoned feeling; a second catalyzes the digestion of poison. The genetic factors that affect the relative efficiency of these two enzymes can play a critical role in whether someone revels in the high from alcohol exposure, or finds it more misery than pleasure. In particular, individuals who have an efficient form of the first enzyme but a slow-acting form of the second face the highest levels of acetaldehyde and are least likely to become alcoholics.

Referring back to our analogy, these are the people who find a roller coaster unpleasant after just a few minutes and are likely to avoid buying a ticket at all. By contrast, individuals who have the opposite chemistry—a slow-acting form of the first enzyme, and a very efficient form of the second—find the alcohol intoxication process quite pleasurable, and want to ride the metaphorical roller coaster as often as possible.

Genetic variants in these enzymes are common and are unevenly distributed around the world. In particular, individuals of Asian background are more likely to be in the first category, and perhaps this accounts for the fact that many Asians report nausea and skin flushing after drinking alcohol, and alcoholism is less common in their part of the world. However, as with all connections between genetic variation and ancestry, this is only a statistical statement about averages; there are certainly plenty of alcoholics in Asia. Similarly, while in general individuals of European background may be less likely to experience the unpleasant side effects of alcohol, there are certainly plenty of people of European descent who are physiologically deterred from alcohol abuse by its more unpleasant consequences.

Tobacco Addiction

Any smoker can tell you that giving up cigarettes is extremely difficult. Nicotine creates a genuine physical addiction. Especially for those who begin smoking early, this addiction creates a deep-seated hunger for tobacco, which is very hard to resist. But just as with alcohol, studies of families and identical twins have demonstrated that the tendency toward tobacco addiction varies between individuals and appears to have significant genetic contributions.

Not only is tobacco addiction under some kind of genetic influence, but it appears that the health consequences of long-term smoking may also vary between individuals, on the basis of genetic inheritance. A particularly interesting recent finding brings these two susceptibilities together in unexpected ways. Three independent groups, attempting to identify why some smokers develop lung cancer whereas others with equivalent exposure to smoking do not, scanned the genome looking for variants that might play a role in susceptibility. All three groups zeroed in on the same part of chromosome 15, where three genes that code for nicotine receptors reside. So the question immediately arose: have these investigators discovered genes for nicotine addiction, or genes that heighten the risk of cancer in people who are addicted for other reasons?

Follow-up studies are conflicting, but there is some chance that both answers may be true. Having two copies of the risk version of these receptor genes apparently increases the risk of addiction, but may also raise the chance of smoking-associated lung cancer in heavy smokers.

Personality Traits

It's one thing to talk about hereditary factors that contribute to cancer, or schizophrenia, or even skin color. But when we start talking about human personality, the information becomes unsettling. After all, we each believe that our identity is based on much more than our DNA. Upbringing and free will should play a central role, perhaps even trumping heredity, when it comes to what kind of people we are. Does not the Declaration of Independence say we are all "created equal"? Equal, yes, but not equivalent.

Have you ever taken a Myers-Briggs personality test? It is one example of a widely applied instrument used to quantify aspects of human personality. The test is intended to provide insight to individuals about how they approach life's problems. And in some instances, employers use the results to assess what kind of performance they might expect from their employees. They do so at their peril, however: these tests are rather crude, often obscuring more than they clarify.

Personality testing has a long and often unhappy history. For example, it was used by the government to weed out homosexuals and other "loyalty risks" during the "red scare." Yet geneticists have reasoned that personality tests might still be useful if they can identify heritable components of human behavior. Twin studies, again, have proved irresistible and fruitful. Robert Cloninger studied personalities in twins and identified seven features of human personality. Four of these appear to be strongly inherited: novelty seeking, harm avoidance, reward dependence, and persistence. Three other personality features could be quantified, but these were far less heritable and appeared to mature only during adulthood: self-directiveness, cooperativeness, and self-transcendence.

Given that heredity seems to be heavily involved in the first four personality components, considerable efforts are under way to identify specific gene variants that might play a role. A considerable buzz arose 10 years ago in the scientific community about a variant in a gene called *DRD4*, involved in dopamine metabolism, which appeared to be associated with novelty seeking. Unfortunately, many follow-up studies failed to confirm this. At best, a recent study identified a related gene (*DRD2*) as associated with novelty seeking—but only in females, and accounting for less than 3 percent of the trait. It seems likely that the genetic contributions to human personality will be numerous, with each one contributing a very small effect. So claims of predictability of personality traits based on DNA testing should be taken with considerable skepticism.

Criminality

When I first arrived at NIH to direct the Human Genome Project in 1993, there was a great furor about a conference planned to discuss genetic factors in criminal behavior. The reasons for this controversy were numerous, and were unfortunately inflamed by ill-considered comments made by a prominent neuroscientist. His remarks seemed to imply, on the basis of no evidence whatsoever, that such genes might be differentially distributed among the races. This racist implication understandably raised intense objections in minority communities. The workshop had to be canceled in order to avoid a major confrontation.

The history of genetics and criminality goes back a long way, and carries regrettable and inflammatory baggage from the eugenics movement in the United States and elsewhere, which gathered considerable momentum in the early twentieth century until it reached its horrendous and genocidal climax in the Holocaust.

The genetics of criminality reared its head again in the late 1960s, at the time when chromosome studies were first becoming possible. A few small reports indicated that some males in penal institutions carried an extra Y chromosome. This XYY condition is now known to occur in approximately one in 1,000 males, but in these early studies the focus was primarily on prisons. The suggestion was made that such a "supermale" might be more likely to exhibit hyperaggressiveness and run into trouble with the law. Subsequent research indicates that these conclusions are unjustified. At the most, there may be a weak association of XYY with mildly reduced intelligence.

A new burst of interest in the genetics of criminality appeared in 1993. That year saw the publication of genetic information on a large Dutch family in which several males showed borderline mental retardation and impulsive aggression, and had committed various criminal acts, including arson, attempted rape, and exhibitionism. All these males were related in a fashion that suggested X-linked inheritance. Sequencing of one particular gene on the X chromosome, *MAOA*, involved in neurotransmitter function in the brain, revealed that all the affected males had a knockout mutation in this gene. Since males have only one X, this meant that they were completely deficient in this function.

Subsequent studies of the *MAOA* gene have further illuminated this story, giving rise to the unfortunate label "warrior gene." The knockout mutation is very rare, but common variations in *MAOA* affect the amount of the enzyme that is produced. A study of males in New Zealand revealed that the variant associated with lower *MAOA* activity was associated with violent behavior and criminal convictions, but only if an individual had also experienced significant maltreatment during childhood. The *MAOA* variant alone, in the absence of child abuse, had no effect. Once again, we see an example of gene-environment interactions in human behavior.

These discoveries about genetics and criminality raise inevitable questions about whether genetically susceptible individuals are still fully responsible for their actions. Could an individual with low *MAOA* activity use "My genes made me do it" as a defense in a criminal court proceeding, seeking leniency or even acquittal? On the other hand, might the presence of genetic susceptibility to criminal behavior cause the courts to be more likely to deliver a longer sentence, reasoning that recidivism might be more likely for such a person? Before rushing to either of these conclusions, we should keep in mind that nearly all genetic factors in criminal behavior will be quite weak in their consequences, so such court decisions based on DNA would be truly ill founded. There is, in fact, a compelling precedent for ignoring such genetic factors in assigning responsibility for moral choices. After all, approximately half of the U.S. population carries a genetic risk factor that places people at a sixteenfold higher likelihood of imprisonment than the other half. That happens to be the Y chromosome. And yet society has not endorsed the idea that simply being a testosterone-intoxicated male is an excuse for committing a criminal act. None of the genetic factors we are now discovering are likely to have nearly as strong an influence on criminality as the Y chromosome.

Male Fidelity

Trumpeter swans form essentially monogamous pair bonds for life. Many other animals, including chimpanzees, engage in multiple promiscuous sexual relationships, with only limited pair bonding. We humans show evidence of evolutionary remnants of both behaviors. But stimulated by the emotional needs arising from higher brain function, the long period of child rearing that requires assistance from both parents, and strong religious traditions and cultural practices, lifelong monogamy of a sexually and emotionally compatible couple is still considered by many the ideal in human relationships. Clearly some individuals are more successful at this goal than others. Among the many factors involved in fidelity, are there genetic variants that might play some role?

Fascinating research on voles, small mammals (related to mice) that can be found in the fields of North America, has shed interesting light on the biological basis of monogamy. Prairie voles make lifelong monogamous pair bonds, whereas their close relatives, the montane and meadow voles, do not, indulging instead in a series of one-night stands. A brain peptide hormone, arginine vasopressin (AVP), acting through its receptor (denoted by the abbreviation V1aR), plays a critical role in this pair bonding behavior. Genetic engineering of variations between these vole species in expression of V1aR can force a profound alteration in mating behavior.

Following up on this observation, investigators have explored natural variations in the human *V1aR* gene, seeking to determine whether these might play some role in male fidelity. Studying more than 500 same-sex twin pairs and their spouses in Sweden, researchers identified a variant in the human *V1aR* gene showing statistically significant association with a variety of measures of marital satisfaction or stress. As just one example, males carrying two copies of the "risk" variant of *V1aR* reported a 34 percent incidence of marital crisis or threat of divorce in the previous year, whereas those carrying no copies of the risk allele reported only 15 percent. Wasting no time, a company in Canada now offers genetic testing of the *V1aR* variants for $99, presumably offering women a chance to check out the possible wandering eye of their prospective mates—or perhaps to provide a wandering male with a biological excuse for his behavior.

Do not be misled, however, by these discoveries. Although the association may be real, and has some scientific interest, the actual influence on the behavior of an individual male is quite modest, and should certainly not be used in mate selection or as an excuse for cheating on one's partner.

Sexual Orientation

Of all the areas where the genetics of human behavior have created controversy, the investigation of genetic influences on homosexuality must rank at the very top. In an appendix to my previous book, *The Language of God*, I cited some scientific data on this topic, not attempting to attach any moral significance to the results, but simply reporting what we know and what we don't know. That brief paragraph has repeatedly been quoted and misquoted in highly inflammatory Internet blogs, often with phrases taken out of context or even with the wording intentionally changed. Each time, the blogger has attempted to propagate one of these two views: (1) homosexuality is completely biologically determined; or (2) there is no biological basis whatsoever for homosexuality, and it is entirely learned, under the control of free will, and therefore reversible.

The facts of the matter lie between these extremes. At this writing, no specific gene variants have been identified that predispose to male or female homosexuality (despite the broad public claims of such a finding

14 years ago). The data from twin studies certainly indicate that such hereditary factors are likely to be found, however, and may be discovered soon. Specifically, if one member of an identical male twin pair is found to be exclusively homosexual, the likelihood that the other twin will also be gay is 20 to 30 percent, depending on the study. This concordance is substantially lower for fraternal twins or for siblings. The baseline incidence of male homosexuality, depending on the study, is in the range of 2 to 4 percent. These facts strongly support the conclusion that hereditary factors play a role in predisposition to male homosexuality, but such factors are not completely determining—otherwise the concordance rate in identical twins would be 100 percent.

There is one other validated finding about male homosexuality that strongly suggests the presence of biological factors. This is the observation that birth order plays a role in male homosexuality. The likelihood of a male's being gay increases by about 30 percent with each older brother. Older sisters and younger brothers have no effect. This has led some people to hypothesize that some maternal immune response against the Y chromosome may in some way affect the sexual development of future male offspring, but there are currently no molecular data to support that hypothesis. Interestingly, if one calculates the proportion of male homosexuals accounted for by this fraternal birth order effect, it may be as much as 30 percent.

Measures of Intelligence

Intelligence is clearly influenced by heredity. Certainly many uncommon genetic disorders like fragile X syndrome (see Chapter 2) have a major effect on intelligence. But what about the role of genes in the general population? One must be careful in drawing conclusions based on any specific IQ test, since clearly such tests are influenced by culture, language, and educational opportunity. However, within relatively homogeneous groups where those external factors are well balanced, it appears that performance on an IQ test reflects approximately 50 percent heredity and approximately 50 percent nonheritable factors. This implies that specific genes will be involved. In a recent study of 6,000 children, however, selected to focus on those in both the lowest and highest range of IQ testing, genome-wide association studies came up almost empty-handed. No single gene variant showed an effect of more than one-quarter of an IQ point. Apparently the genes involved in intelligence are very numerous (perhaps in the hundreds), and each individual variant contributes a very small effect. It is still possible that in the longer term this complicated correlation between genes and intelligence will be sorted out, but one should be very skeptical in the near term of any claims to have found "IQ genes" of any real significance.

Spirituality

It may seem the height of wrongheaded genetic determinism to postulate that genetic influences could play a role in an individual's interest in spiritual matters. Nonetheless, a few years ago a book called *The God Gene*, featured on the cover of *Time* magazine, claimed to have discovered a genetic variant that correlated with self-transcendence. The result was wildly overstated, however; this correlation has never been confirmed, and even if true would have had only an extremely minor effect. In assessing this situation, I can't help reflecting on my own circumstances. I have evolved from being a committed atheist in my twenties to being a firm believer in my fifties. (I describe this journey in *The Language of God*.) During my conversion, there is no evidence that my DNA changed. The notion that spirituality is hardwired cannot be entirely correct.

LESSONS ABOUT GENES AND THE BRAIN

A full understanding of the way the brain functions, including an understanding of consciousness, is still a distant goal for science. The challenges of sorting out the function of the billions of neurons in the brain, and all their trillions of interconnections, makes an understanding of the human genome seem simple by comparison. Our human brains may simply not be sufficiently complex to understand themselves.

It seems likely, however, that genetic factors predisposing to the major mental disorders will emerge in the next few years. These discoveries may allow a much more precise delineation of conditions that are currently lumped together according to descriptive symptoms. The famous physicians' and therapists' tool for establishing diagnoses of psychiatric conditions, *DSM-IV-TR*, bases classification on specific symptoms, but no one working in the field believes that this is optimal. A complete molecular reclassification is critically needed in order to learn how better to diagnose, prevent, and treat these common and serious conditions.

As for disorders with significant environmental contributions, such as situational depression, alcoholism, and nicotine addiction, we can anticipate significant advances in discovering many of the genetic susceptibility factors in the near future. This could be an opportunity to institute preventive measures, since in each instance environmental influences play a critical role in the outcome. This will not be without controversy, however. Would it be useful, for instance, to know at birth if an individual is particularly susceptible to alcoholism or tobacco addiction, in order to assist the child and the parents to be particularly vigilant about not falling into destructive behaviors? Such an opportunity would have to be balanced against the potential of stigmatizing a susceptible person, or even creating a self-fulfilling prophecy.

I am almost certain, however, that complete genome sequencing will become part of newborn screening in the next few years. In that case, this kind of predictive information will become accessible, and if properly handled by both parents and society, could potentially do more good than harm.

As for the nonmedical traits of personality—intelligence, spirituality, fidelity, sexual orientation, etc.—these will remain a topic of considerable recreational interest for humans. We all share great curiosity about such things. But given the multitude of weakly acting genetic factors, and the significant role of the environment and individual free will, DNA analysis probably will be of limited utility. Because of its poor predictive value, such testing will be wildly inappropriate for prenatal diagnosis or newborn screening. For adults, it will make a lot more sense just to measure the trait itself. Despite that, I notice at least one direct-to-consumer genetic testing company that offers to provide a report, at least on a research basis, of human characteristics such as "avoidance of errors," "measures of intelligence," and "memory." Have fun with these tests if you want to, but don't take the results any more seriously than you would any other parlor game.

WHAT YOU CAN DO NOW TO JOIN THE PERSONALIZED MEDICINE REVOLUTION

1. Researchers are beginning to sort out the complex patterns of gene expression in the human brain. One particular project, the Allen Brain Atlas (supported by Paul Allen, a cofounder of Microsoft), has been systematically cataloging these patterns, initially in the laboratory mouse but more recently in the human brain. For fun, have a look at http://www.brain-map.org. You might try clicking on "human cortex" and then entering "MAPT" as the gene name for your search (MAPT, the microtubule-associated protein tau, plays a major role in Alzheimer's disease). The table that comes up will list all of the human brain sections available where MAPT expression has been determined. For starters, click on image section 80561119 (from donor 2898, a 35-year-old male), and you will see four sections of the brain. The blue staining shows where the *MAPT* gene is being expressed. Click on any one of the sections to see a higher-resolution image. You can move around using the "pan" buttons, and zoom in to see the microscopic details, rather like with Google Earth. With the fully zoomed-in version, the dark blue triangular structures you see are individual neurons.

2. The "bible" of clinical classification of mental disorders in the United States, published by the American Psychiatric Association, is the *Diagnostic and Statistical Manual of Mental Disorders*

(DSM). The current edition, *DSM-IV-TR*, was published in 2000, and runs to almost 1,000 pages. You can read more about this at http://en.wikipedia.org/wiki/Diagnostic_and_Statistical_Manual_of_Mental_Disorders. Yet this classification scheme is based almost entirely on subjective evaluation of symptoms and signs, and different professionals often arrive at different diagnoses for the same patient. The advent of molecular classification of mental disorders is likely to turn *DSM* upside down in the future.

CHAPTER EIGHT

Genes and Aging

If you happen to carry genes conveying an 80 percent chance that you will suffer from Alzheimer's disease by age 85, would you want to be told? Many people would say no to this question, since at present there is no proven medical intervention to stave off the onset. Yet for many of those who have chosen to be tested and have discovered such a high risk, they recover quickly from the impact of the result and embrace life all the more, while they still can. As we consider the many recent discoveries about aging and our genes, keep that thought in mind. What information would allow you to live your current life to the fullest?

Meg Casey was just three and a half feet tall, but at 23 years of age she could curse like a sailor and had proved to be a legendary advocate for disability rights in her hometown, Milford, Connecticut. As a genetics fellow at Yale, I was both honored and intimidated to be assigned her care as part of my training, but I grew to admire this diminutive figure greatly during the three years that I took care of her.

Meg was afflicted with a disorder of accelerated aging. Her appearance was that of an extremely aged individual, with leathery mottled skin, severe osteoporosis, and very little hair underneath the outlandish wigs she wore. Yet given the diagnosis she carried, Hutchinson-Gilford progeria syndrome (HGPS), she was an outlier, having survived considerably longer than the typical life span of 12 or 13 years for this condition.

Progeria is incredibly rare, affecting no more than 1 in 4 million live births. Progeria seems to strike like a bolt from the blue. Meg's situation was typical in this way, in that there was no family history and none of her six brothers were affected.

At the time I took care of Meg, very little research had been undertaken on this condition. So I was at a loss to understand the progressive process that was inexorably robbing Meg of her physical capacities and that would ultimately take her life a few years later. As a budding genetics researcher, I filed away the study of progeria in the back of my mind, as something that might be amenable someday to molecular understanding. But in 1984 I had no idea about how to proceed.

Sixteen years later, at one of those ubiquitous Washington receptions, I encountered a young pediatric emergency physician who was serving as a White House fellow. I was stunned to learn that his son, now aged four, had recently been diagnosed with Hutchinson-Gilford progeria. He and his wife, a physician-scientist, were understandably distraught by the grim prognosis for this disorder, and by the absence of any significant research effort to understand how it might be treated.

I met their son Sam a few months later. He had already lost all of his scalp hair, and his skin was showing signs of aging. But like Meg Casey, he was smart, spunky, and full of energy and determination. I agreed to help his parents seek ways to encourage further research on this condition. But soon I became drawn into that effort myself, assigning the search for the progeria gene to a new postdoctoral fellow in my laboratory. This was a crazy project to undertake, as none of the tricks that a geneticist would normally use to identify the general location of the responsible gene were available. The disease almost never recurs in families. There were no clues about where to hunt for the culprit in the genome.

The search for the cause of progeria took the postdoc and me down many weird and unexpected pathways, ultimately representing almost an entire course in human genetics. And in less than a year, the search led to the discovery of a single mutated letter in the DNA code, a T that should have been a C, located in the middle of a gene that codes for the protein lamin A.

We had access to DNA samples from just 25 patients with progeria—samples carefully banked over many years by other investigators, in hopes that this discovery would someday be possible. Nearly all these samples showed exactly the same mutation from C to T in lamin A, although DNA from the parents was always normal. In other words, this was a new mutation in each case (geneticists call this *de novo*). By indirect means we were able to show that the misspelling nearly always occurred in the sperm. (This also explained why progeria was somewhat more common in the children of older fathers, where sperm have gone through more cell divisions, and have a greater opportunity to pick up mistakes.) It was breathtaking to consider that this one single-letter mutation, out of a 3 billion-letter genome, was the cause of such a dramatic disease.

However, in our collection were a few DNA samples that did not show this mutation. For one of those, we found two different misspellings in the same lamin A gene, and noted that this individual had had a somewhat unusual course, living longer than the usual patient with progeria. As I scanned the clinical description of this sample, the hair stood up on the back of my neck. There at the bottom of the form, in the space for the name of the submitting researcher, I found my own signature. This was a DNA sample from Meg Casey. I had forgotten that I had obtained her permission to send this sample off to the DNA bank 20 years earlier, in hopes that someday it might be useful to someone. Meg was gone, but she was still helping.

THE BIOCHEMISTRY OF PROGERIA

Because biochemists and cell biologists had studied the lamin A protein for many years, we could almost immediately predict why this mutation from C to T was causing such havoc. Lamin A provides a major component of the structure of the nucleus of the cell, keeping it in its elegant oval shape. Lamin A also has to help the nucleus come apart and then re-form again, each time the cell divides. To accomplish its role, lamin A includes a fancy signal in the tail of its complex structure, allowing it to be properly targeted to the nucleus. Think of that signal as a zip code tag, if you will. On arriving at its destination, that tag needs to be removed, so that the protein can perform its normal functions. What we discovered was that the progeria mutation prevents removal of the zip code tag. To understand why this matters, imagine a bunch of kids who have to get to school on their bicycles. The bikes are critical, or the students won't make it to school. But it is also critical for the bikes to be left in the bike racks outside, and not brought into the classroom, or havoc will result. In progeria, there are too many bikes in the classroom, and normal schoolwork becomes impossible.

What we decided to do was, in effect, ask some of the students to walk to class instead of bringing their bikes. To do this, a drug that had already been developed for another purpose was available. The drug seemed to work nicely in progeria cells growing in a petri dish, but would it work in affected kids? And would it be safe?

WHY DO WE AGE, ANYWAY?

We'll come back to Sam and the story of his dramatic aging disorder, but first some background on aging. Through the centuries, we have been searching for the fountain of youth, trying to stave off the inexorable march of the aging process. We dream of ways to avoid the steady loss of physical capabilities, and to rewrite the end of the play captured in Shakespeare's immortal words: "The last scene of all, that ends this strange eventful history, second childishness and mere oblivion, sans teeth, sans eyes, sans taste, sans everything."

Why must this be so? Is aging an inevitable outcome for all living things? As far as we can tell, bacteria seem to have an unlimited life span, if they are given the appropriate nutrients to live upon. Yet for more complicated organisms, limitations appear. One can imagine two possible reasons for this:

1. Running down of the system is inevitable. For a complex multicellular organism, every time the genome is copied the potential for mistakes creeps in. As organisms age, more and more of those mistakes accumulate in various cells of the body. Likewise, the proteins that do the work of the cell can acquire damage due to environmental exposures or simple occasional acts of random misfolding. The result is an increasing accumulation of nonfunctional or even toxic proteins.

2. Evolution doesn't "want" organisms to live forever. The success of natural selection depends on natural variation and lots of reproduction. If older generations live too long, they may compete for precious resources. So evolution would tend to favor biological upgrades that enhance reproductive success during earlier phases of the life span of an organism—even if, or maybe especially if, this results in a limitation of total life span.

This can't be taken too far, however. A purely evolutionary explanation for limited life span in humans seems to fly in the face of the existence of menopause in females, since this routinely limits the reproductive contributions of women to only a portion of their maximum life span. Interestingly, there are considerable data to suggest that this circumstance can be explained by the "grandmother effect": the presence of older but infertile women provides experienced support to younger parents, increasing the overall reproductive success of the family.

LESSONS ON AGING FROM ANIMALS

The study of yeast, worms, flies, and mice has yielded some surprising discoveries. Many people had predicted that the process of aging was so ubiquitous, and affected by so many interlocking pathways, that no single gene could possibly have more than a tiny effect. However, at least in these model organisms, it is possible to find single genes that can affect the life span by as much as fivefold. Imagine if it were possible to suddenly expand from our life span to age 500! Even Methuselah, with his biblical life span of 969 years, might be within reach. But a close reading of the reports on model organisms suggest that it will be quite a while before these observations can be applied to humans. Here are some key findings from this research.

RESTRICTING CALORIES IS REALLY IMPORTANT

In roundworms exposed to famine, the organism can go into a low metabolic state. Upon recovery from that "hibernation," the worm experiences a longer life span, essentially making up for lost reproductive time. That same system may be a universal theme in animals, though it has yet to be definitively proved in humans. It seems that environmental changes that reduce intake of calories, and thereby reduce insulin signaling, are capable of extending life span, so long as the restriction is not so severe as to result in malnutrition. From studies on baker's yeast, it appears that a particular class of genes may play a master role in mediating this effect. Specifically, these genes are naturally up-regulated in the presence of reduced calorie intake, and that seems to play a central role in the aging process. Furthermore, if these genes can be artificially up-regulated by other means, such as recombinant DNA, the life span is extended. Is this the genetic fountain of youth?

Great excitement has greeted the idea that stimulating the protein products of these genes, called "sirtuins," might provide a pathway toward increased longevity. Interestingly, one of the first compounds discovered that activates sirtuins is a naturally occurring molecule, resveratrol, which is present in red wine and has been credited with the slight reduction in heart disease that individuals who drink a glass of red wine each day seem to enjoy. But resveratrol is a relatively weak activator of sirtuins, so biotechnology companies are seeking to prepare modified versions with considerably greater potency. Two such drugs are now in clinical trials. But once again, the U.S. regulatory system presents an obstacle. Since the FDA does not have a category for longevity drugs, these drugs are being evaluated for their ability to prevent diabetes and heart disease, with a possible "side effect" of longer life!

But it's a bit early to cancel your life insurance. Not all scientists agree that sirtuins are the central mediator of the normal aging process. And currently it is not clear whether the data on the benefit of calorie restriction in other animals will turn out to apply to humans. Finally, even if they do apply, very few humans can tolerate a 30 percent reduction in normal calorie intake over long periods of time. In essence, we may face a choice: eat only 1,200 calories per day and maybe live longer or: eat, drink, and be merry, for tomorrow we die. It isn't an easy decision to make.

DNA INTEGRITY

If you want your cells to remain healthy, their DNA instruction book has to be carefully protected. Any errors that creep in might be expected to accelerate the aging process. In Hutchinson-Gilford progeria, for instance, it is likely that the disruption of the cell nucleus by the lamin A mutation results in progressive DNA damage as cells divide and then have difficulty reassembling their nuclei. Other rare disorders of premature aging, with names like Werner's syndrome and Cockayne syndrome, turn out to be caused by mutations in the DNA repair machinery, further emphasizing the importance of DNA integrity.

A particularly important feature of genome health during aging is the special machinery that is required to keep the *telomeres*—the tips of chromosomes—from fraying. These telomeres can be thought of as analogous to aglets—those plastic or metal caps that keep the ends of your shoelaces from getting ratty. Chromosomes use an enzyme called telomerase instead of plastic. The DNA sequence of all human telomeres is a long repeat of six letters: TTAGGG. Without repair, the length of the TTAGGG repeat gradually shrinks each time the cell divides. As telomeres get shorter and shorter, the genome of that cell is in greater and greater jeopardy, and ultimately a signal is triggered that causes the cell to commit suicide. Telomerase is the enzyme that re-extends the repeat sequence, preventing the cell's demise. Interestingly, *stem cells* have plenty of telomerase, and therefore can keep dividing almost indefinitely. Similarly, cancer cells almost always turn out to have activated their telomerase genes. But most other cells of the human body that are destined to age and die are no longer producing this enzyme, and therefore have a limited ability to reproduce themselves before the system runs down.

IS LIFE SPAN HERITABLE?

We often comment on families with long individual life spans by saying they must have "good genes." Is there any truth to that idea? I will admit that I hope the answer to this question is yes, as both of my parents lived to be 98 years old. My maternal great-grandfather lived to 105, and practiced law until the age of 100. Of course, if I crash my motorcycle into a tree next year, my genetic endowment may not save me from an early death. But assuming that I dodge the vagaries of external accidents or epidemics, should I expect to reach age 100?

The answer is maybe. Studies of families and identical twins reveal that about 20 to 30 percent of an individual's life span appears to stem from heredity. However, if one restricts this analysis to individuals who have made it to age 70, then the evidence for heritability gets much stronger. Apparently this means that for people who escape a variety of causes for early death, their longevity rests more significantly on their genes than on the environment.

The effort to identify specific genes that are involved in maximum individual life span is just beginning to get under way, and so far there are few solid clues. My own laboratory has identified some common variations in the lamin A gene that may play a small but reproducible role in life span. Undoubtedly, there will be more of these common variations that turn out to have a small but statistically significant effect. Like depression, perhaps, aging is probably controlled by a number of different genes. The discovery of their variants may not allow anyone to predict your life span, but may point toward important pathways involved in the normal process of aging. That in turn may shed light on ways to improve the chances of healthy aging for more and more people.

The study of telomeres will undoubtedly continue to contribute to our understanding of the aging process in humans. In fact, there is already evidence in some studies that telomere lengths in circulating adult white blood cells correlates with expected life span. Longer telomeres correlate with longer survival. A recent study of Swedish twins indicates that these longer telomeres reflect both genetics and environment. Most striking of all, but still somewhat controversial, is a recent observation that individuals who score high on an optimism scale have longer telomere lengths. Whether this can be extrapolated to say that lifelong optimism provides an actual biological mechanism toward extending life span cannot be proved by existing data, but is an interesting speculation. Can you, in effect, hope your way to a longer life?

STEALING THOSE GOLDEN YEARS: ALZHEIMER'S DISEASE

Four and a half million Americans are affected with Alzheimer's disease. So it is perhaps surprising that this brain disorder was not described until 1906. Perhaps that is a reflection of the progressive increase in average life span during the twentieth century: more and more of us now live long enough to get this terrible disease. Alzheimer's disease usually begins after age 60. It is an irreversible and progressive brain disease that slowly destroys memory and intellectual function. The pathological hallmark of the disease under the microscope is the appearance of abnormal clumps of material (amyloid plaques) and tangled bundles of fibers (neurofibrillary tangles). The plaques turn out to be made up primarily of a peptide, amyloid-beta (Aβ).

About 5 percent of individuals with Alzheimer's disease have an earlier onset (in their forties or fifties). In those cases there is often a strong family history of early-onset disease. It turns out that most of these families have mutations either in the amyloid gene or one of the enzymes that processes amyloid to Aβ.

For the remaining 95 percent of cases, heredity is still important, though the pattern is less clear. Studies of families and identical twins reveal that roughly 70 percent of the risk of late-onset Alzheimer's disease is genetic. A major component of this hereditary influence derives from variations in a gene called *APOE,* which has three common variants in the human population.

It turns out that the "ε4" allele of *APOE* is a significant risk factor for Alzheimer's disease. Specifically, a person who carries one copy of ε4 has a threefold increased risk of Alzheimer's disease, and someone carrying two copies faces an eightfold increased risk. Translated into absolute numbers, that means a person with one copy of ε4 has a 30 percent chance of developing Alzheimer's disease by age 85, compared with the 10 percent chance in the general population (see Figure 8.1). And the risk is even higher for the double-copy carriers, approaching 80 percent.

However, *APOE* is not the whole story—there must be other genetic and environmental risk factors that contribute to this distressing disorder. One possible contributor may be head trauma. Wartime head injuries causing unconsciousness have been implicated. Boxers who have sustained repeated blows to the head as part of their profession sometimes develop early-onset dementia, referred to as "dementia pugilistica." Boxers with the same "bad" variant of *APOE* seem particularly predisposed to this outcome. Perhaps genetic testing should be offered to young men before they decide to adopt this profession.

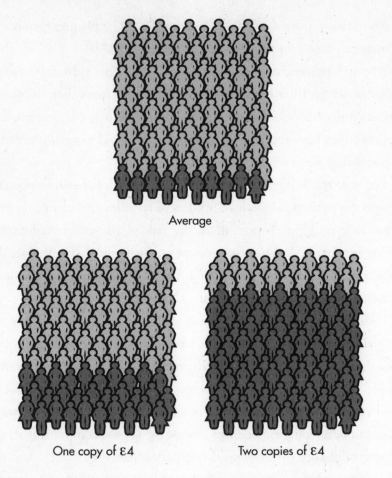

Figure 8.1: Risk of Alzheimer's disease by age 85, depending on *APOE* genotype.

More generally, given the high frequency of Alzheimer's disease, the possibility of offering genetic testing for *APOE* has been debated for almost 20 years. Would you want to know? Let's recall the RBI (risk × burden × intervention) formula from Chapter 3. The risk R, associated with the bad variant ε4, is significant and well established. The burden B, for individuals and their family, of Alzheimer's disease is unquestionably profound. But what about the I intervention factor? A few years ago, there was considerable interest in the possibility that statins (designed to lower cholesterol) might also reduce the risk of Alzheimer's disease—but that has not been a consistent finding in later studies. There has also been a suggestion that vigorous mental exercise (crossword puzzles and sudoku are often mentioned) might delay the onset—but that, too, is far from clear. Apart from avoiding head trauma, therefore, there are currently no proven interventions available to prevent or even delay the disease in a susceptible individual. The medical treatment of Alzheimer's disease is generally unsatisfactory, and the limited drug therapies available do not seem to work differentially for people of different *APOE* genotypes. So in terms of effective interventions, the I factor in the RBI equation is currently not very impressive.

Nonetheless, some individuals are genuinely interested in their future risk of Alzheimer's disease, and wish to have the opportunity to plan ahead. For them, the ability to plan should probably also be considered as part of the I factor. Should they be denied the chance to find out their genetic risk? Will some of them be devastated by the results? In an effort to determine whether *APOE* testing provides information that individuals can utilize in a positive way, a large-scale research study has been under way for ten years: the Risk Evaluation and Education for Alzheimer's Disease Study (REVEAL). The participants are adult children of patients with Alzheimer's disease, so they have been highly sensitized to the nature of the condition, and are concerned about their own future. Volunteer participants in RE-VEAL were randomly assigned to groups that would or would not receive information about the *APOE* genotype, and then were followed over the course of one year to assess their response to the disclosure of the information. They were informed ahead of time that there were no proven medical benefits to learning their *APOE* genotype. After testing, the members of the "disclosure" group were given their DNA results, and told their risk of Alzheimer's disease. In this carefully controlled study, there was no evidence that those at increased risk experienced undue anxiety relative to those who were not. Interestingly, however, the ε4 carriers *were* more likely to adopt changes in health behavior. They took vitamins, dieted, and exercised, even though they had been told that none of these interventions has been proved to prevent or delay Alzheimer's disease.

I spoke to one of the participants in the REVEAL study. Mark (not his real name) was 67, and he had watched Alzheimer's disease steal the final years from his mother, his aunt, and his uncle. So he was not really surprised to find that he carried a copy of *APOE* ε4, nor was he surprised at the roughly 30 percent risk he was quoted. A retired physicist, he was familiar with statistics, and knew that this also meant there was a 70 percent chance he would not be affected. He told his physician about the result, but asked that it not be placed in his medical record. In searching for actions that he might take to reduce his risk, Mark heard that statins might potentially slow the development of Alzheimer's disease. He was already taking a statin for his mildly elevated cholesterol, and on the basis of the *APOE* result, he consulted with his physician and doubled the dose. (Note that this is controversial—could he be hurting himself?) He also made a decision to schedule some international travel to New Zealand and Switzerland that he might otherwise have put off. And he continues to watch the listing of clinical trials at www.clinicaltrials.gov (a Web-based compendium, operated by the National Institutes of Health, listing all clinical trials being conducted on human diseases), to see if any new research studies are being conducted on preventing Alzheimer's disease.

So Mark certainly used the genetic information to make some life decisions, though not all of them were scientifically proved (doubling his statin dose, for instance, was unproved). He did not think he had suffered psychologically from getting this news, though he does find that he watches his own mental performance more carefully now. If given the chance to do it all over again, he would still choose to know his status.

Does this mean that testing for Alzheimer's should be made available to the general public? It should be noted that REVEAL involved intensive genetic counseling and educational steps prior to and following testing, and considered only a single condition and a single genetic test. Furthermore, REVEAL focused specifically on individuals with an affected parent, who were more likely to be knowledgeable about the disease.

What about you? Do you want to know your risk of Alzheimer's? Much of the answer comes down to whether you think planning for the future is a sufficient reason to know your risk. Interestingly, three of the first individuals to have had their complete genomes sequenced—Craig Venter, James Watson, and Steven Pinker—have all wrestled with this issue. Watson and Pinker chose not to learn about their *APOE* status, though they released the rest of the DNA sequence to the public. Venter did the genomic "full Monty," including *APOE*. He turned out to be at increased risk.

When I recently underwent genetic testing, the *APOE* result caused me the greatest anxiety. Although fully aware of a negative family history, was I ready to learn about this particular risk factor, with all of its potential consequences? I knew that this DNA test was not destiny, just a measure of predisposition, but I wondered how a positive result would change my view of the future. I strongly considered just ignoring this part of the report. But ultimately curiosity got the best of me. Finding no copies of ε4 was a relief. But of course I could still develop Alzheimer's disease—it's just the likelihood is reduced, down to about one chance in 30.

SLOWING DOWN AGING?

The tongue-in-cheek *Journal of Irreproducible Results* once published a paper titled "The Genetics of Death," concluding that death follows dominant inheritance with 100 percent frequency of the death allele. And of course this is true: the certainty of death and taxes seems secure. Though some futurists may forecast an era of regenerative medicine that allows repair of virtually all body tissues, it is probably best to assume that the death rate will continue to be one per person. What then can you do to maximize your threescore and ten, perhaps to even stretch it to fivescore or more?

The advice basically falls into two main categories.

First of all, do everything you can to avoid preventable chronic illnesses that may both shorten life and reduce its quality. Avoiding smoking, following a diet that is well balanced, engaging in regular exercise, taking a low-dose aspirin a day (for men), avoiding extreme sun exposure, and having regular medical evaluations to look for early signs of treatable disease, influenced by your own family medical history, are all activities that will promote an extension of healthy life.

Second, for those highly motivated individuals interested in trying promising but unproven approaches, consider a few other activities that may retard the aging process. On that list would be reduction in caloric intake to approximately 30 percent less than the normal requirement. This represents an austere lifestyle that many will find difficult to sustain, but may well turn out to have longevity benefits in the long run. Also, pay attention to progress with new drugs based upon the principle of stimulating the action of sirtuins. These may find their way into general use in the next few years, and this is certainly an area to watch. Meanwhile, a glass (not the whole bottle) of red wine now and then is both pleasant and potentially life-extending.

Other approaches that aim to slow aging by limiting "oxidative stress," as with large doses of vitamin E, are certainly heavily advocated in health food stores, but should be viewed with some skepticism in the absence of more data to support them (there are even some data that suggest harm). Ultimately, the development of additional pharmaceutical approaches to reduce DNA damage, support telomere maintenance, and allow repair of tissue that has been damaged by aging may well become practical. But this is certainly an area where consumers need to be skeptical. Ponce de Leon never found the fountain of youth, and it is unlikely to turn up in the next few years.

WHAT ABOUT SAM?

We may be philosophical about the slow progression of the normal aging process, especially when we are not feeling its effects, but Sam and his parents have no such luxury. The average age at death of a child with progeria is 12, the cause almost always being a heart attack or a stroke. As I write this, Sam is 12 years old. Is his clock running out? Maybe. But maybe not. There is now substantial hope for Sam and other children with this condition.

You may recall that we had identified a drug that seemed capable of reducing the amount of toxic protein in progeria cells growing in the laboratory. In terms of the metaphor used earlier in the chapter, the drug persuaded many of the students to walk to school instead of riding their bikes. This drug was subsequently tested on a mouse model of progeria that my laboratory had created. Untreated, the mice had the same cardiovascular problems that take the lives of most kids with progeria. But when the drug that had worked well in lab studies, a farnesyl transferase inhibitor (FTI), was given to the mice, the cardiovascular disease was prevented. To our amazement, the FTI drug could even reverse the cardiovascular damage after the mice had been left untreated for several months.

With this evidence, and the apparent limited toxicity of the drug, a clinical trial for 29 children with progeria was started in 2007. Sam was one of the first to be enrolled. Two years into the trial, Sam appears to be doing well. Although it is very difficult to know whether the drug is actually reducing his risk of heart attack or stroke, and it may take several years to find out, there is great hope that Sam may benefit from this experimental strategy.

You might wonder whether any of these discoveries about the genetics and biochemistry of progeria are relevant to the normal process of aging. Yes, children with progeria seem to age at about seven times the normal rate, but does that relationship to aging extend to the molecular level, or is this a purely superficial relationship? Interestingly, in the last two or three years it has been possible to show that all of us make small amounts of the same toxic protein product from the lamin A gene that causes progeria in children. This is a somewhat disturbing finding. As we age, we all have some bicycles in our classroom! The toxic protein becomes more abundant in our cells, and is easily detectable in the elderly. It may well be that the triggering of production of this protein is a significant contributor to the limitation of the normal human life span. Therefore, what we learn about progeria may well have direct implications for all of us. It is a bit soon to contemplate putting the FTI drug in the water supply, since this drug may have many other long-term side effects in normal people, but the strategy of studying a rare and dramatic form of aging may well teach us important features of the normal process. One is reminded of a famous remark by William Harvey in 1657, advocating for the importance of studying rare conditions: "Nature is nowhere accustomed more openly to display her secret mysteries than in cases where she shows tracings of her workings apart from the beaten path;

nor is there any better way to advance the proper practice of medicine than to give our minds to the discovery of the usual law of Nature by careful investigation of cases of rare forms of diseases. For it has been found in almost all things, that what they contain of useful or applicable nature is hardly perceived unless we are deprived of them, or they become deranged in some way."

CONCLUSION

The growing evidence that heredity has a role in virtually all diseases, and even in the normal process of aging, is becoming ever more apparent. With the new tools of genomics, this knowledge becomes not only academically interesting but potentially actionable. But what if an individual, despite all these opportunities for prevention, still becomes ill and requires a drug treatment? Surely then the decision about the proper drug and the proper dose could be made without recourse to DNA analysis? Not at all—this area of therapeutic medicine promises to be among the very first where personalizing overturns the "one size fits all" approach.

WHAT YOU CAN DO NOW TO JOIN THE
PERSONALIZED MEDICINE REVOLUTION

1. Several Web-based tools are available that aim to estimate your "biological" age—as opposed to your calendar age—based on your personal health habits and medical history. Some of these will also make recommendations about how to improve your score, though not all of those are backed up by rigorous data. For a look at this information, try the RealAge tool at http://www.realage.com/ral ong/entry4.aspx?cbr=GGLE626&gclid=CJKh8Pal_ ZkCFeRM5QodKk0iGQ and see how you come out.

2. The National Institute on Aging of the National Institutes of Health (NIH) provides useful information about healthy aging. Check out http://www .nia.nih.gov/HealthInformation/.

3. The Centers for Disease Control and Prevention (CDC) also provides materials on healthy aging, including the chance to enroll in an e-mail list that provides new information as it comes to light. You can enroll at http://www.cdc.gov/aging/.

CHAPTER NINE

The Right Drug at the
Right Dose for the Right Person

Something was terribly wrong with McKenzie. She had always been a happy and energetic little girl, but now at age 12 she had lost her appetite, suffered from unexplained stomachaches, and seemed listless. After several visits to doctors, the awful truth finally emerged—McKenzie had acute lymphocytic leukemia (ALL).

Deeply worried, McKenzie's parents took her to the Mayo Clinic. They were encouraged to hear from the pediatric oncology experts that combination chemotherapy now cures 85 to 90 percent of kids with ALL. But the side effects of these powerful drugs can be severe. McKenzie gritted her teeth and prepared for loss of hair, nausea, fatigue, weight gain, and susceptibility to infection.

But McKenzie actually just missed a serious brush with death, and not only because of the leukemia. Had she been given the standard doses of drugs for ALL, she might have died from the treatment. Fortunately for her, the Mayo Clinic was one of the few places in the world in 2000 that were testing kids for their ability to handle one of the powerful chemotherapeutic agents used for treatment of ALL, a drug called 6-mercaptopurine (6-MP). Genetic researchers at the Mayo Clinic and St. Jude's Hospital had learned a few years earlier

231

that about one in 300 individuals is missing an enzyme for metabolizing 6-MP. Given a standard dose of this drug for age and body weight, such individuals will build up dangerously toxic levels, which suppress their bone marrow and expose them to potentially fatal risks of infection, bleeding, or both.

McKenzie turned out to be one of those rare kids missing the enzyme. Since the Mayo doctors knew this ahead of time, thanks to a simple blood test, they were able to adjust her treatment plan. She was still able to receive 6-MP, but less than one-fifth of the standard dose. She and her parents recalled having to cut a very small pill into even smaller pieces, and wondering how such a tiny shard of a tablet could do any good.

But McKenzie did beautifully. Her leukemia went into remission within weeks, and she tolerated the two years of drug treatment extremely well. She was even able to stay in school throughout almost all of her course of therapy. Today she is 21 years old and has no sign of the disease.

WHY DON'T DRUGS ALWAYS WORK THE WAY THEY SHOULD?

A major source of frustration for doctors and patients alike is that the outcome of drug therapy is not always what one hopes for. If 100 patients are properly diagnosed with a particular condition and given the standard dose of the most appropriate approved medication, on average 70 or 80 of them will benefit. The others will not, and a few may even suffer a toxic reaction. The proportions that fall into each of these classes will depend on the specific drug, but virtually no drug gets perfect marks.

It is bad enough if the drug provides no benefit, but when the result is a toxic side effect, then a sacred principle of medical ethics, "First, do no harm," has been violated. Regrettably, this happens every day. A recent study estimated that in the United States each year, more than 2 million hospitalized patients suffer serious adverse drug reactions, with more than 100,000 of those resulting in a fatal outcome. Adverse drug reactions are the fifth leading cause of death in the United States. Adverse reactions that occur in outpatients are undoubtedly even more common, but these come to official attention only if the attending physician submits a report to the FDA. That system is entirely voluntary. Given a society that tends to be very intolerant of bad outcomes from drug therapy, it is shocking that there is no systematic network to collect data about such events.

WHAT ARE THE CAUSES OF ADVERSE DRUG OUTCOMES?

A myriad of reasons can contribute to the high incidence of undesirable outcomes from drug therapy. An astonishing number come about simply because of poor handwriting: the doctor writes an illegible prescription, and the pharmacist provides the wrong drug or the wrong dose to the unsuspecting patient. If ever there was a mandate for computerized record keeping and electronic medical records, this is it! Also, instructions about the dose and how often the drug should be given are often confused in the transmission of information from health care provider to patient. That problem is made substantially worse where many different medications are being taken each day. In short, the biggest reason for potentially deadly drug reactions is simple human error. But this isn't the only reason. Underlying illnesses, particularly of liver or kidneys, can have a substantial impact on the way drugs are metabolized and cleared from the body, and can readily result in adverse outcomes if impairment of those organ systems is not recognized and accommodated.

Other problems can arise when multiple drugs are being given and interactions between them occur, so that the effect of one or more of the drugs is much stronger or weaker than expected. Many such interactions have been cataloged, and many pharmacies are set up to recognize the potential for trouble when a new prescription is written, but such interactions still slip through all too often.

And even if all of these potential confounders of proper prescribing were always addressed, the variation in individual response to drug therapy would *still* be substantial. If you have made it this far in the book, you will not be surprised to learn that much of the basis of that variation lies in DNA. The study of how drug response is influenced by the genome is called *pharmacogenomics*.

HOW CAN YOUR GENES AFFECT YOUR DRUG RESPONSE?

In order to understand how genetic variation can influence drug response, it is helpful to consider the typical steps involved in drug action. They are depicted in Figure 9.1. First, some drugs are not actually provided in their active form, but must be converted by an enzyme (A) to a compound that is biologically capable of producing the desired effect. Next, a metabolic tug-of-war begins, as other enzymes (B) in the body seek to degrade the active form of the drug, converting it to an inactive substance, which is ultimately excreted. The active form of the drug meanwhile must interact with some biological component of the body, often a receptor (C), to induce the desired outcome. Since the instructions for each of the three steps in Figure 9.1 (A, B, and C) are encoded by genes, and since most genes have some common variants, it is not surprising that individuals differ in their reactions to standard doses of a drug.

Drug metabolizing enzymes, depicted as A and B in Figure 9.1, are often but not always located in the liver. It is not always possible to predict whether someone with a relatively low-activity version of one of these enzymes will need a higher or a lower dose of the drug, since you need to know whether the usual form of the drug is its active form, or the pro-drug form. A person who is a slow metabolizer of a pro-drug (enzyme A) will produce an inadequate amount of the active form, and therefore will probably have limited benefit. On the other hand, a slow metabolizer of the active form of the same drug (enzyme B) may experience toxicity for that drug.

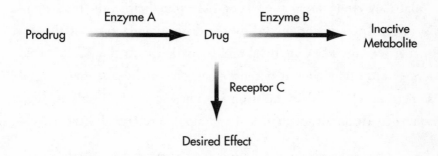

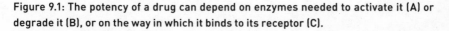

Figure 9.1: The potency of a drug can depend on enzymes needed to activate it (A) or degrade it (B), or on the way in which it binds to its receptor (C).

Not only can the enzymes involved in drug metabolism harbor variants that can affect drug response; the receptors (C) upon which the drugs act can also vary between individuals. Clearly, if a receptor has a very poor affinity for a drug, the result will be disappointing. A receptor with exceptionally high affinity can produce a toxic response, even at normal blood levels of the drug.

A BRIEF ASIDE ON NAMING DRUGS

When I entered medical school in 1973, I was not excited by the need to memorize a lot of material—I hoped that medicine would be all about general principles. It came as a bit of a shock, therefore, that

I could not really learn how to practice medicine without learning the names of hundreds of different drugs, as well as understanding their mode of action and their proper administration. My distress was multiplied because all drugs that are approved for general use in the United States essentially have *two* names, and a health care provider needs to be generally familiar with both. One of these names, often referred to as the generic name, tells you some small bit of information about the compound (for instance, all generic names that end in "-mab" are *m*onoclonal *a*nti*b*odies). The other name, the trade name, is assigned to the compound by the drug manufacturer at the time that it receives FDA approval and is marketed to the public. Trade names are often chosen because they are catchy and may suggest something about what the drug is supposed to do. Table 9.1 gives some generic and trade names of commonly prescribed drugs.

Table 9.1—Examples of Commonly Prescribed Drugs

Generic Name	Trade Name	Purpose or Target
atorvastatin	Lipitor	Lower cholesterol
clopidogrel	Plavix	Avoid clotting
esomeprazale	Nexium	Reflux
imatinib	Gleevec	Leukemia
levofloxacin	Levaquin	Infection
paroxetine	Paxil	Anxiety, depression
sildenafil	Viagra	Erectile dysfunction
trastuzumab	Herceptin	Breast cancer

PREDICTING THE RIGHT DOSE

A major opportunity for pharmacogenomics will be in understanding how knowledge of genetic variation can be used to get the dose right for a specific individual. Doctors have tried to do this in the past by taking into account such things as age, gender, body weight, possible interactions with other drugs, and the presence of liver or kidney disease, but that has not always led to a successful outcome. We can do better.

6-Mercaptopurine in Childhood Leukemia

We began this chapter with the story of McKenzie, who developed ALL—a childhood leukemia—and was in need of aggressive chemotherapy to cure it. McKenzie had a nonfunctioning variant of the enzyme thiopurine methyltransferase (TPMT), which serves as enzyme B in Figure 9.1, normally inactivating 6-mercaptopurine so that it can be excreted. The recognition of this condition made it possible for McKenzie to receive the drug, but at a much lower dose. Oddly enough, though this information has been available for some time, the FDA still does not require TPMT testing before administration of 6-mercaptopurine.

Although ALL is a relatively rare disease, a number of other, related drugs are also metabolized by TPMT, so these findings have broader implications. For example, the drug azathioprine is often used for severe cases of rheumatoid arthritis, and individuals with poorly functioning copies of TPMT can experience toxicity from this drug if the dose is not properly adjusted.

Clopidogrel (Plavix)

A major advance in the treatment and prevention of heart attacks (coronary artery disease) has been the availability of drugs that block the formation of clots in the arterial system. Such clots generally are initiated by the aggregation of blood platelets, and so compounds that inhibit such aggregation can be of considerable benefit. In fact, low doses of aspirin achieve this goal, and are recommended for healthy men over age 40 with one or more risk factors for heart disease, and likewise for healthy women over age 65. A more potent platelet aggregation inhibitor is the drug Plavix (clopidogrel). It is clear, however, that individual response to this frequently prescribed drug is not uniform, and recent evidence suggests that the enzyme CYP2C19 may explain a fair amount of this variability. Plavix is a pro-drug, and so individuals with low-function CYP2C19 obtain less benefit. It seems likely that this variability in response might be overcome by administering higher doses, but that is currently under study.

Antidepressants

Drug treatment of clinical depression has made great strides in the last few decades. However, initiating such therapy can be a frustrating experience, as typically it takes several weeks to see a response. Many patients fail to respond to the first drug chosen, so that other alternatives must be tried empirically, often prolonging the agony of depression for months.

To facilitate the choice of the right drug for each individual, without all this trial and error, genetic variations are being sought that might be good predictors of the optimal choice of drug and dose. But this is a highly complex area of medical research. Not all cases of clinical depression are the same. Other life events may act to worsen or improve depression, and even achieving a standard definition of drug response is not simple. Therefore, at least at the time of this writing, the effort to utilize genetic analysis to optimize the treatment of depression has not yet reached the point of effective implementation.

Coumadin (Warfarin)

Many observers believe that coumadin will be the first drug for which the so-called Dx-Rx paradigm—a genetic test (Dx) followed by a prescription (Rx)—will enter mainstream medical practice. This drug has a long and checkered history. In the 1920s, farmers in Wisconsin were distressed to observe an epidemic of hemorrhage in cattle. Severe bleeding would occur from a minor injury, or even spontaneously. Investigation of the possible cause led to a batch of moldy silage made from sweet clover, which had been fed to the affected animals.

Twenty years later a chemist working at the University of Wisconsin was able to purify the compound in the silage that caused the bleeding. It was realized that this might be a valuable treatment, if properly administered at the right dose, to prevent blood clots in humans. The compound was given the trade name Warfarin, in recognition of the Wisconsin Alumni Research Foundation, which had supported the research.

President Eisenhower was an early recipient of the drug in 1955, after his heart attack, and gradually the indications for administration of coumadin extended to the prevention of deep vein thrombosis in the legs, as well as the prevention of strokes in individuals with atrial fibrillation, an abnormal heart rhythm.

In 2004, 31 million prescriptions were written for coumadin. But it was also listed among the 10 drugs with the largest number of serious adverse events reported that year. On U.S. death certificates, anticoagulants, primarily represented by coumadin, rank number one in deaths from adverse drug effects. Those deaths often come about because of inadvertent overdoses, leading to intestinal hemorrhage or bleeding into the brain. No fewer than 29,000 visits to emergency rooms each year are attributed to coumadin overdoses.

Trying to achieve the proper dose of coumadin is a nightmare for both physician and patient. The dose must be closely monitored by a blood test that checks the actual effect on blood-clotting factors, and the initiation of therapy is always a white-knuckle experience. Underdosing may expose patients to the risk of clotting just at their most vulnerable time, and overdosing can result in a serious or even fatal hemorrhage. But individuals vary by as much as tenfold in the dose needed to achieve the desired therapeutic effect. Some of this variability can be predicted on the basis of age, gender, body mass index, and smoking history, but more than half of the variability cannot be predicted on the basis of these variables.

It was therefore of great interest to discover that common variations in two genes, CYP2C9 and VKORC1, can account for about 40 percent of the variability in therapeutic dose (see Figure 9.2). The first of these, CYP2C9, codes for an enzyme that metabolizes coumadin (enzyme B in Figure 9.1), so underactive versions of CYP2C9 can lead to toxicity unless the daily amount of drug is cut back. The second, VKORC1, has its effect in a different way, which can be understood only by reference to the mechanism by which coumadin carries out its anticlotting activity. Specifically, as shown in Figure 9.3, coumadin blocks a step needed to activate four blood-clotting factors. That activation step also requires vitamin K. But vitamin K needs to be regenerated after each use, and the enzyme VKORC1 is responsible for this. In individuals who have a slower-acting form of VKORC1, the relative insufficiency of vitamin K slows the activation of the blood-clotting factors, and so a smaller dose of coumadin is needed in order to achieve the same effect. This is a more complicated mechanism than what is depicted in Figure 9.1, and yet it demonstrates the many ways complex pathways in drug metabolism and body biochemistry can affect the proper dose of a drug.

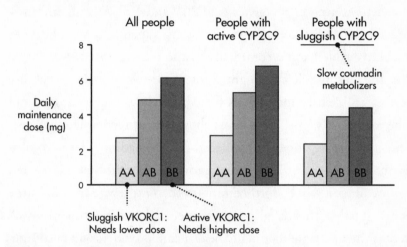

Figure 9.2: Variants in the *CYP2C9* and *VKORC1* genes play a substantial role in determining the optimum maintenance dose of coumadin. The *VKORC1* gene has two different forms: the A form is less active, the B form more active. Individuals are either AA, AB, or BB. (Data taken from M. Rieder, et al. *New England Journal of Medicine* 352 [2005]: 2285–93.)

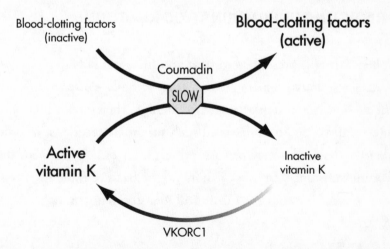

Figure 9.3: Vitamin K is critical for activating a number of blood-clotting factors; coumadin blocks that step. But vitamin K itself needs to be reactivated after each use, and that is carried out by the enzyme VKORC1. Individuals with sluggish VKORC1 will thus need less coumadin to achieve a particular level of blood thinning.

The results of the study shown in Figure 9.2 were retrospective (based on DNA testing of a study conducted earlier). But in February 2009, researchers reported results of a large prospective study showing that genetic testing of the genes *CYP2C9* and *VKORC1* significantly improved the ability to achieve a stable therapeutic dose. The FDA had already added some comments to the coumadin label in 2008, suggesting that physicians might want to be aware of the possibility of testing to identify individuals for whom the standard dose might not be ideal. This label stops short, however, of recommending testing for all who are about to be placed on coumadin. A larger trial is now under way, and it is the expectation of many of us that such a recommendation will be forthcoming in the next few years. Given the very broad use of this drug in the practice of medicine, this may well represent the first time that many health care providers will encounter a requirement to follow the Dx-Rx paradigm.

PREDICTING UNCOMMON TOXIC REACTIONS

When a drug is prescribed to treat an illness but fails to give benefits, that is a serious problem. If the drug actually causes a serious side effect, that is a potential tragedy. There is, therefore, a high level of interest in trying to understand such rare toxic reactions, in order to identify those who may be susceptible and avoid giving them the offending substance. In a few instances, the basis of these reactions has been worked out, but the future will identify many more.

Abacavir (Ziagen)

In the 1990s, an intense effort to identify drugs active against HIV resulted in rapid FDA approval of several of these compounds.

Among them was abacavir, a drug designed to block replication of the AIDS virus. Abacavir is highly active against HIV, but a major problem soon appeared. Specifically, about 6 percent of those given the drug developed a dramatic hypersensitivity reaction, consisting of fever, skin rash, gastrointestinal symptoms, and respiratory distress. If the drug was not stopped, the hypersensitivity could progress to a very severe state, and could even be fatal.

A search of the genome to uncover the genetic basis for this hypersensitivity reaction turned up a very straightforward answer. Virtually all the risk of hypersensitivity to abacavir relates to a genetic variant on chromosome 6, located within the HLA region. HLA refers to a gene cluster that codes for many proteins involved in the immune response, and individuals who carried a particular variant, HLA-B*5701, turned out to be the ones susceptible to hypersensitivity. In 2004, a study in Australia demonstrated that avoiding the use of abacavir in patients who were HLA-B*5701-positive almost completely eliminated the hypersensitivity reaction. But the rigorous system for evaluating data in the United States forced the initiation of another large study of almost 2,000 patients. When this was published, it confirmed that hypersensitivity was essentially limited to those positive for HLA-B*5701. In July 2008, the FDA added information to the drug label, recommending that individuals should be screened for this genetic variant before abacavir is prescribed. This is a particularly simple example of applied pharmacogenomics, where a single gene is highly predictive of a toxic effect.

Statins

This class of drugs includes the best-selling medications in history. Highly effective in lowering blood cholesterol, these medicines are now taken by millions of people, and have been credited with greatly improving long-term survival of individuals who have had a previous heart attack or who have demonstrated coronary artery disease. There is still some controversy about whether individuals with high cholesterol but no demonstrated heart disease also experience prolongation of life by taking statins, but the weight of evidence suggests that they do.

But if you happen to be one of those 2 out of every 100 individuals who develop weakness and muscle pain at the higher doses of statins currently prescribed, you may not be so enamored of this particular drug class. Although that side effect is generally reversible, it can be temporarily quite distressing, and the biochemical measurements done on the blood of individuals who experience it indicate that there has been substantial muscle damage.

Given that 98 percent of people taking statins, even at relatively high doses, do not experience this side effect, it was natural to investigate whether some genetic variation might account for susceptibility. Very recently, a fascinating answer has appeared. A single gene, *SLCO1B1*, has emerged that accounts for a substantial fraction of the risk of muscle side effects from statins. Those who carry one copy of a common variant in *SLCO1B1* have a fourfold increase in the risk of muscle toxicity, and those who carry two copies increase that risk by a factor of 16. The biological explanation in this instance is fairly clear: *SLCO1B1* encodes a transporter in the liver that mediates uptake of statin drugs. This is relatively new information, and the fact that muscle toxicity from statins is reversible makes identifying susceptible individuals potentially less critical, but it seems likely that in the future those who carry the double copy of the *SLCO1B1* variant will be discouraged from taking statins, or perhaps will need a much lower dose.

PREDICTING LACK OF DRUG RESPONSE AT ANY DOSE

There is no point in administering a drug to an individual who can be reliably predicted to be unresponsive. Use of the drug in such a case exposes the person to possible toxicity, wastes money, and deprives the individual of access to an alternative therapy that might have been more successful. But in the past, most failures to respond were inexplicable and unpredictable, and could be identified only after the fact. That is all beginning to change, as the following examples will show.

Herceptin

The first widely relevant example where a genetic test (Dx) is required before a drug is administered (Rx) is the use of Herceptin (trastuzumab) for the treatment of breast cancer. Twenty years ago, it was discovered that breast cancers in some women expressed a molecule on the surface of their cancer cells that might be a useful drug target. This molecule, called HER2, is a receptor for growth factors, and therefore plays a significant role in the ability of the cancer to grow. Researchers reasoned that blocking the receptor would provide an advantage in cancer treatment, and so a monoclonal antibody (trastuzumab) against HER2 was generated. Testing in clinical trials showed substantial benefit for the treatment of breast cancer, but only for cancers that expressed HER2. As a result, the FDA now requires tumor cells from a woman's breast cancer to be tested to see whether HER2 is present, and Herceptin is to be prescribed only if the test is positive.

There have been some problems with this implementation of Dx-Rx, however. The genetic test for HER2 is difficult to perform, and results are not always entirely reliable. Furthermore, despite the FDA's recommendations, doctors don't always follow the rules. A major insurer recently reported that 8 percent of women who were treated with Herceptin had actually tested negative for HER2, and an additional 4 percent had not been tested at all.

It is important to note that the HER2 test cannot be done on DNA from blood, saliva, or a cheek swab. Instead, the test must be done on the cancer cells themselves, as the expression of HER2 is a reflection of mutations that have occurred during the development of cancer. This should not be confused with an assessment of hereditary risks. One can anticipate many more examples of this sort of genetic testing of tumor samples to assess the appropriateness of intervention. Another one was described in the case of Karen Vance in Chapter 1, and in the cases of Judy Orem, Marvin Frazier, and Kate Robbins in Chapter 4. Another recent example is the realization that two new therapeutic monoclonal antibodies—cetuximab (Erbitux) and panitumumab (Vectibix)—directed against the growth factor receptor EGFR, do not seem to provide any benefit to patients whose tumors carry an activating mutation in the oncogene *KRAS*. This may at first seem anomalous, in that the cancer cells may still have EGFR on their surface. But it turns out that KRAS is "downstream" of EGFR in the signaling pathway. If KRAS is mutated and locked in the "on" position, it won't matter whether EGFR is shut down by the drug or not. To put it another way, pulling the driver's foot off the accelerator won't slow the car down if the accelerator is already stuck.

Tamoxifen

It is not just experimental drugs for which genetically determined differences in response rates can be found. Tamoxifen is an estrogen antagonist, and for more than 30 years has been a mainstay for the treatment of breast cancers that are positive for the estrogen receptor. And yet, as with virtually all drugs, not all women treated with tamoxifen have experienced a good outcome. Until recently, that was assumed to be a matter of different biological behavior of the cancers: some cancers were said to be too aggressive to be suppressed by this hormonal treatment. But new information just coming to light challenges that explanation. Tamoxifen is actually a pro-drug (see Figure 9.1) and must be converted by the enzyme CYP2D6 to its active form, called endoxifen. Individuals with genetic variations in *CYP2D6* that result in low activity of this enzyme thus do not receive the benefit of a standard tamoxifen dose, and may be poor candidates for this particular strategy. No definite recommendations have yet been issued, but a woman being given this drug might wish to inquire about whether she is likely to be a responder, on the basis of this new information about genetic variations in *CYP2D6*.

OBSTACLES TO THE PHARMACOGENOMICS REVOLUTION

In this chapter we have discussed almost a dozen examples where there is good evidence that genetic information can have a clinically useful impact on drug response. Given that optimizing outcomes for drug therapy ought to be a high priority, and especially given the high proportion of adverse drug reactions and drug failures, why hasn't the pharmacogenomic approach already moved into the mainstream?

First, conducting convincing studies to identify genetic factors associated with rare adverse reactions to a drug can be challenging. If the toxic response strikes only one in 1,000 patients, then doing a study that enrolls 20,000 treated individuals will still identify only 20 cases. That will often be insufficient to identify the cause. What is desperately needed here is an effective system to capture reports of adverse drug reactions once a drug has been marketed, when millions of people may be taking it. Coupled with the ability to obtain blood samples from the affected individuals, such a system could lead to rapid discovery of the cause. The lack of such an effective reporting system is a major flaw in the United States, and should be remedied as quickly as possible.

Second, when the question being asked is whether individuals in a certain subset—those with a particular diagnosis—are just not likely to respond to a therapy, the study design should be somewhat simpler—but here the motivation to do the study may be limited because of economics. A drug manufacturer is unlikely to be interested in conducting a study that will reduce the size of the market for a drug the company has just developed at great cost. Thus, such studies are likely to be done only if the cost of the study is borne by another organization, such as the National Institutes of Health.

Third, the regulatory agency, which in the United States is the FDA, appears reluctant to require genetic testing for the administration of any particular drug, unless compelling data have been produced by several independent studies. Despite strong evidence, no such recommendation has been made for the use of 6-MP in childhood leukemia, and the lack of a requirement for such testing may be exposing children like McKenzie to the possibility of severe drug toxicity. As another example, the evidence suggesting the need for genetic testing to avoid hypersensitivity reactions to abacavir was apparent five years ago, but only after a second large prospective trial did the FDA put the requirement for testing on the label.

Fourth, health care providers may be slow to take advantage of the Dx-Rx paradigm, even in circumstances where the data are compelling and the FDA has indicated that testing is required. As noted above, even with Herceptin, for which the requirement has been in place for more than a decade, 12 percent of patients are not benefitting from the personalizing of decision making that would optimize their care.

Fifth, logistics can be a barrier. Although with some drugs for chronic illness a few days' delay can tolerated so as to wait for the results of a genetic test, in many other circumstances the drug needs to be started immediately. At present, nearly all pharmacogenomic testing is done in centralized laboratories. This means that samples must be shipped to and analyzed in a central lab, and then the results must be conveyed back to the point of care. Several days can easily elapse before the result is known. That is simply unworkable for many applications of drug therapy. Short-term solutions to this problem will require optimizing the shipping process, or exporting the most commonly used genetic tests to hospital laboratories and doctors' offices. But such solutions will often be unsatisfactory. Part of this problem is also traceable to an unwillingness of third-party payers to reimburse for the test.

A more definitive solution will be to obtain the complete genomic sequence of each of us, to put that information into our medical records, and to make it available for inquiry when needed. When the cost of such sequencing becomes affordable (probably no more than five years from now), it will be much more cost-effective to obtain the DNA information up front, making this information immediately accessible at moments of medical need. The mandate to implement pharmacogenomics in medical practice will therefore be one of the strongest drivers for the coming era of complete genomic sequencing for all individuals.

Prediction of future disease risk, optimizing of medical therapy, and discovery of new approaches to treatment of disease are all in the midst of upheaval. The transformation of medical care is already breathtaking. But now, let's stretch beyond the current reality and imagine where medicine may go in the future. It promises to be a wild ride.

WHAT YOU CAN DO NOW TO JOIN THE PERSONALIZED MEDICINE REVOLUTION

1. If you have already taken advantage of one of the direct-to-consumer services providing large-scale genetic analysis (as described in Chapter 3), look specifically at your own results that relate to pharmacogenomics. Does your report include those? If not, contact the company and ask whether that information can be provided.

2. If you haven't done this, or if the service you used did not provide that information, you might want to consider obtaining this information on yourself to guide future drug prescriptions. One possible source of testing is the company DNA Direct at http://www.dnadirect.com/web/consumers. Have a look a the information it provides about "Testing for Drug Response."

3. Are you taking prescription or over-the-counter medications now? Are you sure that these do not represent a risk of adverse drug interactions? Most pharmacies check for this with every new prescription, but not everyone gets all their prescriptions from the same pharmacy, and over-the-counter drugs generally aren't tracked this way. There are several Web-based tools available for free that will allow you to check for such drug interactions. Have a look at http://www.healthline.com/druginteractions for one that is easy to use.

CHAPTER TEN

A Vision for the Future

When I got off the plane in Lyon, France, in February 2001, I was delighted to see that I had landed at the Antoine de Saint-Exupéry airport. One of my favorite books when I was a child was Saint-Exupéry's *The Little Prince*. Many truths about life and love were interwoven in this strange little story of the prince who lived on an asteroid.

I was in Lyon to give a presentation describing what we had learned from the first draft of the human genome sequence. But I had also been asked to make some predictions about the future.

This kind of crystal ball gazing has not always gone well. In 1943, Thomas Watson, the chairman of IBM, said, "I think there is a world market for maybe five computers." A few years later, a professor at Yale evaluating a student's hypothetical business plan wrote, "The concept is interesting and well formed, but in order to earn better than a C, the idea must be feasible." This student was Fred Smith, who proposed forming a company that became FedEx. And in 1962, Decca Records rejected a quartet's demo, saying, "We don't like their sound, and guitar music is on the way out." The group was the Beatles.

Yet none other than Saint-Exupéry once said, "As for the future, your task is not to foresee, but to enable it." Having read out a draft of the human DNA instruction book, the Human Genome Project was certainly positioned to do some enabling. So I tried to base my predictions on that foundation.

What predictions did I make? These six for 2010:

1. Predictive genetic tests will be available for a dozen conditions.
2. Interventions to reduce risk will be available for several of these.
3. Many primary care providers will begin to practice genetic medicine.
4. Preimplantation diagnosis of fertilized eggs will be widely available, and its limits will be fiercely debated.
5. A ban on genetic discrimination will be in place in the United States.
6. Access to genetic medicine will remain inequitable, especially in the developing world.

My audience at Lyon thought predictions 1 to 5 were too bold. Many skeptical comments were voiced. But as I write this now, on the brink of 2010, it is fair to say that I was actually too conservative. If I had predicted the coming of direct-to-consumer DNA testing, for modest fees, I would have been hooted off the stage.

It's time now for the next set of predictions. Where is personalized medicine going in the coming decades?

But before answering that, we need first to consider two radical new approaches to therapeutics that we have so far barely touched on. The true promise of both strategies remains controversial, but if successfully applied they could result in truly dramatic breakthroughs in the treatment of a long list of diseases.

GENE THERAPY

Hopes for *gene therapy* have waxed and waned for two decades. In the 1990s, they first surfaced for extreme cases of immunodeficiency. David Vetter, the famous "boy in a bubble," lived from 1971 to 1983, had no immune system, and could never come into direct contact with anything unsterilized. Finally, in despair, his care givers tried an experimental bone marrow transplant, even though no perfect match could be found. At first, the transplant appeared to be successful, but then David developed a form of cancer that turned out to be due to a virus present in the donor marrow. As the world watched and grieved, David slipped away. On his gravestone are these words: "He never touched the world, but the world was touched by him."

In medical terminology David's disease is severe combined immunodeficiency (SCID). The type that he suffered from is referred to as *X-linked* SCID, and affects only males. David had an older brother who died in infancy, and this fact helped alert the doctors to David's condition.

Six years after David's death, another dramatic scene unfolded at the National Institutes of Health, where Ashanthi DeSilva, a four-year-old girl affected by a different type of SCID due to mutations in a gene called *ADA*, underwent the first human experiment in gene therapy.

The goal was to insert a normal copy of the *ADA* gene into a sufficient number of her immune cells so that they would recover their function and serve effectively to protect her against infections.

Gene therapy can be accomplished in some instances by removing cells from the body, adding the missing DNA, and then returning them to the patient. This is generally referred to as *ex vivo* gene therapy, and is diagramed in Figure 10.1. But many body tissues (such as the brain or the heart) cannot be removed and replaced in this way, and so other gene therapy strategies depend on direct administration of the genes into the human body. This is referred to as *in vivo* gene therapy.

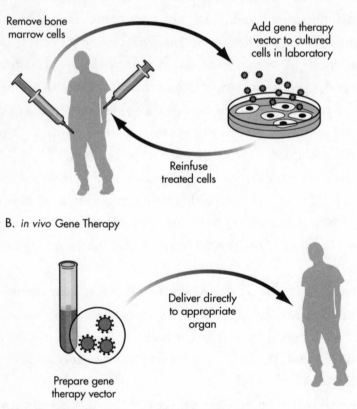

A. *ex vivo* Gene Therapy

Remove bone marrow cells

Add gene therapy vector to cultured cells in laboratory

Reinfuse treated cells

B. *in vivo* Gene Therapy

Deliver directly to appropriate organ

Prepare gene therapy vector

Figure 10.1: *Ex vivo* and *in vivo* approaches to gene therapy.

There were three reasons why ADA deficiency was chosen for the first human experiment in gene therapy: (1) It was a potentially fatal disease, so the unknown risks were worth braving. (2) The cells of the immune system could be harvested from the bone marrow, making possible the *ex vivo* approach, which seemed simpler and safer for the first application of gene therapy. (3) It was predicted that any immune cells that were successfully corrected by the insertion of a normal copy of the *ADA* gene would have a selective growth advantage over uncorrected cells. Therefore, the corrected cells would be expected to outcompete their deficient neighbors, and might allow reconstitution of a complete immune system, even if the gene therapy process was rather inefficient.

Prior to this experiment, many debates had been held about whether it was ethically appropriate to transfer foreign genes into the cells of a human, even if the intentions were benevolent. Was this "playing God"? But the ethical use of organ transplantation had been accepted for several decades, and transplanting a heart or a kidney certainly represents a much more wholesale transfer of foreign genes than the insertion of a single therapeutic gene. A residual concern revolved around whether there was any possibility that the foreign gene would actually be transmitted to future generations, with unknown consequences. But for every gene therapy protocol ever attempted in humans, the likelihood that the foreign DNA would end up in the sperm or eggs of the treated individual has been extremely low.

Ashanthi tolerated her gene therapy quite well. There appeared to be some improvement in her immune function, but retreatment was necessary. Apparently the cells receiving the *ADA* gene were not as long-lived as hoped. Complicating the picture, a new drug therapy for her disease had just been developed, and for compelling ethical reasons it was tried on her at the same time. She is now in college and doing well, still receiving the drug treatment. It remains hard to say just how much her gene therapy mattered.

Despite this uncertain outcome of the first trial, gene therapy seemed poised to revolutionize medicine in the early 1990s. Many bright young clinical investigators joined the effort to develop appropriate applications. However, researchers' early enthusiasm gradually faded, because of three huge challenges:

1. **Delivery.** DNA is a large, charged molecule, and to get it to travel across the cell membrane and the nuclear membrane is not an easy task. Furthermore, if the therapy is to be effective, a substantial fraction of the cells in any target organ must receive the foreign DNA. Investigators have therefore focused on the use of naturally occurring viruses, which after all have been optimized by evolution to succeed at this exact delivery task. The viruses are first inactivated so that they cannot cause disease themselves, and then harnessed as delivery trucks to transfer the desired DNA sequence to the nucleus of the appropriate target cells. All this has proved to be difficult.

2. **Function.** Getting the DNA into the appropriate cells does no good unless it is actually transcribed into RNA, and then translated into the desired protein product. If the viral vector is designed to integrate itself into one of the human chromosomes (as opposed to existing as a free-floating piece of DNA), then the actual site of integration is crucial. Just as with planting a seed, the hospitality of the landscape matters. Sometimes, the new DNA is successfully inserted, but then immediately shut down by its neighbors. If the DNA does not integrate into a chromosome, on the other hand, it is likely that over the course of time it will be diluted by cell divisions.

3. **Immune Response.** Viruses provide an efficient means of DNA delivery, but also code for foreign proteins that the immune system is quick to recognize. Despite major efforts to reengineer viruses so that they can evade recognition, the immune system is often more clever than the human engineers. Just at the point where some benefit from the gene therapy starts to appear, the immune system often seeks out and destroys the cells expressing the therapeutic gene.

Despite those obstacles, researchers pressed on. Then, in 1999, tragedy struck. An 18-year-old volunteer in an *in vivo* gene therapy experiment to treat a missing enzyme in the liver died suddenly, just three days after being infused with the therapeutic virus. The young man, Jesse Gelsinger, apparently died from a massive activation of his immune system in response to this foreign substance.

A detailed investigation of the circumstances revealed that certain safeguards had not been fully adhered to. Worse yet, the principal investigator of the study appeared to have a potential conflict of interest: his involvement in a biotechnology company. It was the end of innocence for gene therapy researchers. They were inured to frustration, but they had never expected to do any real harm.

After some serious regrouping, a small band of investigators has pressed on. Over the last few years, they have become newly optimistic, though not without additional setbacks. Once again, children with SCID have been at the forefront of these second-generation gene therapy approaches. Twenty boys with X-linked SCID, the same disorder that killed David Vetter 20 years earlier, were treated with a gene therapy. Initial reports of reconstitution of their immune systems led to rejoicing in the community. That enthusiasm was short-lived, however, as 5 of these 20 patients unexpectedly developed a form of leukemia. This unfortunate outcome was most likely a direct consequence of the gene therapy, as the insertion of the virus had activated an oncogene.

Fortunately, all but one of those five boys have been successfully treated for their leukemia. So despite this setback, for boys with X-linked SCID who lack a matched sibling donor for a bone marrow transplant, gene therapy provides a better likelihood of long-term survival than any other alternative. David Vetter would probably be alive now if this approach had been available in the 1970s.

The news has recently become even more encouraging. A recent Italian report of 10 boys and girls with ADA deficiency (the same disease that Ashanthi has), all lacking a suitable bone marrow donor, indicates that eight of them appear to have been cured by gene therapy, without any evidence of ill effects.

But will gene therapy remain applicable only to rare disorders of the immune system? What about applications to other diseases? To answer that, meet Dale Turner.

Dale appeared normal in every way in early childhood, but by the time he was five it became apparent that he wasn't seeing well. After multiple visits to specialists, he was diagnosed with a condition called Leber's congenital amaurosis (LCA). This is a classic example of a medical term that tends to obscure reality; amaurosis means blindness! The disorder LCA is a recessive disease, due to mutations in a gene called *RPE65*. No treatment was available.

As time progressed, Dale had more and more difficulty living normally. He was able to enroll in college, but he needed access to special technology that could scan printed material and turn it into spoken words.

In November 2005 Dale was stunned to get a letter from his doctor, asking if he would be interested in participating in a gene therapy trial for his disease. Experiments with a dog version of LCA had shown promise, and it was time to initiate human trials.

Dale took two months to decide whether to enroll in the gene therapy trial; the consent forms were scary, and he was afraid that something might go wrong. But on the last day of January 2008, he was wheeled into an operating room for gene therapy. A virus was injected into a small sector at the back of his right eye. The experience was unnerving, especially when he heard the surgeon say in the midst of the procedure, "I've got a retinal tear here; can you pass me the laser?" But the operation was nonetheless judged a technical success.

Dale waited to see whether there would be any benefit. For the first three days he was told to stay inside, but on the fourth day he went outdoors, wearing sunglasses. He could not resist taking a peek at the sky. He was astounded to see how blue it was, and he realized how muted and drab colors had become to him. Over the next few weeks, the vision in his right eye became better and better. He noticed that he could close his left eye and see his nose! Simple things like the clarity of the fine detail in a field of grass or the wood grain of a table were a source of great exhilaration.

Tests (Figure 10.2) demonstrated that the sensitivity of the treated part of Dale's retina had improved more than 100-fold. That improvement has persisted without any evidence of deterioration for more than a year. Buoyed by this improvement in vision, Dale entered law school in the fall of 2008. He finds that he can actually keep up quite well with the reading assignments, without depending on special equipment. Treatment of the other eye is already planned.

Dale is a remarkable success story, and he's not the only one. Gene therapy treatments for patients with LCA have now been reported in both the United States and the United Kingdom. Although the results are not uniformly successful, they show great promise. Unquestionably, however, gene therapy has been on a roller coaster for the last 20 years, with lots of disappointing plunges. The successes in treating immune deficiencies, LCA, and a few other conditions are exceptional. But could there be an alternative to virus delivery that changes the game significantly?

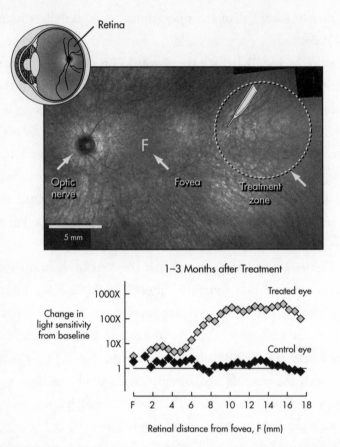

Figure 10.2: Dramatic improvement in Dale Turner's vision is apparent after gene ther-apy for his visual disorder. The area of the retina treated is marked by the dashed circle in the photograph of the back of Dale's eye. The graph shows a more than hundredfold improvement in visual acuity in the treated eye. (Clinical data kindly provided by Drs. Artur Cideciyan and Samuel Jacobson of the University of Pennsylvania)

THE COMING REVOLUTION IN STEM CELL THERAPY

Indeed there could. If delivery of genes to cells is potentially beneficial but technically challenging, why not just deliver the cells? After all, that's what we do in organ transplantation—although the shortage of organs and the serious threat of rejection have limited this approach. The recent recognition of the unexpected ability to reprogram human cells has opened up an entirely new vista of potential therapeutics. To understand this, we need to delve a bit into developmental biology, and we need to consider the concept of stem cells.

Essentially, a *stem cell* is one that is capable of self-renewal, and also of reprogramming (scientists call this "differentiating") into other cell types after an appropriate stimulus. There are many kinds of stem cells. The mother of all stem cells is the single-cell embryo created by fusion of sperm and egg, carrying within it the potential to implant into the uterus and give rise to a live-born child. Such a stem cell is called "totipotent." As the human embryo divides into two cells, and then four, and then eight, this total potential remains only briefly. By the time 100 cells have appeared, these cells have differentiated into some that are destined to form the body of the fetus, and others that make up the placenta.

While the spherical embryo is still only the size of a pencil dot, cells in a certain group (called the "inner cell mass") remain capable of differentiating into all the tissues of the human body, though they now lack the potential for forming a placenta. These cells are called "pluripotent." Experiments with mice over several decades indicated that these cells can be grown indefinitely in the laboratory and maintain their pluripotency.

What about adults? No totipotent or pluripotent stem cells apparently remain in an adult human, but other types of stem cells do in fact exist in various body compartments, most notably the bone marrow. These cells, though, are more limited in their capacity to differentiate into multiple cell types, and are therefore called "multipotent." The existence of adult stem cells makes it possible to transplant marrow from a donor to a recipient, repopulating the entire blood-forming and immune systems of the recipient by multipotent stem cells contributed by the donor. There is even a suggestion that adult bone marrow stem cells may be able to differentiate into other tissues, and perhaps contribute to the repair of heart muscle after a heart attack.

The more limited capacity of multipotent cells for self-renewal and differentiation makes it unlikely that they will provide therapeutic solutions for many diseases for which treatments are badly needed. In the lingo of the profession, pluripotency is much more powerful than multipotency. That's why embryonic stem cells are such a hot topic. Their almost unlimited potential for becoming any desired cell type has created a great deal of scientific interest. It has also created an enormous uproar in bioethics. The standard method of preparing human embryonic stem cells (Figure 10.3A), first worked out just 10 years ago, is to remove cells from the inner cell mass of a human embryo that has been produced by in vitro fertilization (IVF). With careful culturing in the laboratory, these cells can grow almost indefinitely, in which case they are called a "stem cell line."

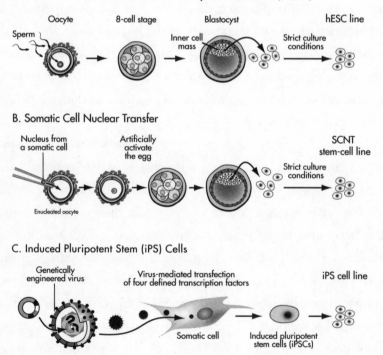

A. Traditional Derivation of Human Embryonic Stem Cells (hESCs)

B. Somatic Cell Nuclear Transfer

C. Induced Pluripotent Stem (iPS) Cells

Figure 10.3: Three different types of human stem cells, created in very different ways.

For those who believe passionately that human life begins at the moment of conception, the use of a human embryo for research is deeply troubling. However, the IVF process, considered by many an acceptable means of assisted conception, generally produces several embryos. Implanting too many of these into the woman's uterus makes a multiple pregnancy highly likely, with serious risks to both mother and offspring. As a result, hundreds of thousands of human embryos have been frozen and stored in IVF clinics.

It is inevitable that the vast majority of those frozen embryos will ultimately be discarded. Is it more ethical to destroy them this way, or to use a small number of them to generate new human embryonic stem cell lines that might be useful in treating terrible disorders like spinal cord injury, diabetes, or Parkinson's disease? That debate has raged in the United States. In 2001, President Bush decided to allow federal funding of research using human embryonic stem cell lines that had been derived by August 9 of that year, but not for any lines developed after that. This rather arbitrary deadline satisfied very few people, and was overturned by President Obama in 2009.

I share the conviction that the product of sperm and egg is a potential human being, and should be treated with dignity and respect. In that regard, I am deeply troubled by the creation of human embryos solely for research purposes. I am reassured, therefore, to see that the Obama administration has chosen to disallow federal funding of stem lines derived in this way. (Full disclosure: I served on the part of the Obama transition team that deliberated extensively on stem cell policies.) But if adequate consent is obtained from the parents, and no inappropriate incentives are involved, the research use of excess embryos produced during IVF, otherwise doomed to be discarded, seems to me to be a defensible ethical position—especially because there is also an ethical mandate to try to help people who suffer from otherwise untreatable conditions.

In January 2009, the first human trial of human embryonic stem cells was approved for the treatment of spinal cord injury. All too often, when people injure their spine (in a car crash, for example), they become paralyzed throughout the body, below the location of the injury. Other methods of trying to reverse such paralysis have generally been unsuccessful. Recent experiments on rats have shown that stem cells can encourage reconnection of the transected spinal cord. This first human trial is not actually intended to assess effectiveness; it is meant primarily to look for any unanticipated toxicity. Still, it represents a milestone.

Once one begins to contemplate the potential of stem cells to replace or repair tissue, the number of conditions that we might treat in this way begins to grow. Diabetes, in which the insulin-producing cells of the pancreas are too few in number, and Parkinson's disease, in which a specific group of neurons in the brain are dying, are just two such targets. Yet researchers face several practical problems that will sound familiar from our discussion of gene therapy. Delivery of the cells to the appropriate location can be a challenge; getting them to perform the appropriate functions in that location is difficult; and the threat of rejection by the immune system looms if the cells are derived from another individual.

Two very exciting recent developments suggest that there may be a potential solution to the problem of rejection. If pluripotent cells could be derived from the patients themselves, then presumably no immune suppression would be needed at all. Until recently, such an idea seemed far out of reach. But that was before Dolly the sheep ambled onto the front pages of the world press.

Dolly was produced by a procedure called somatic cell nuclear transfer (SCNT). In this procedure (Figure 10.3B), a cell from a mature animal (often a skin cell, but for Dolly it was an udder cell), containing the entire genome of that animal, is persuaded to travel back in time and recover its totipotency. This amazing trick is accomplished by taking the nucleus of the skin or udder cell and inserting it into an egg cell from which the nucleus has been removed. Signals from the cytoplasm of that egg cell are somehow capable of reprogramming the nuclear DNA. In the case of Dolly, that cell was implanted into the uterus of an unrelated sheep, and gave rise to a genetic clone of the original donor.

This process of so-called reproductive *cloning* electrified the world 10 years ago, and has been successfully carried out since then for cows, horses, cats, and dogs. It is not clear, however, whether this reprogramming to totipotency will work in humans; an initial claim by a Korean group of success in producing a totipotent human cell by SCNT was later revealed to be a fraud.

The profound ethical concerns about SCNT relate to the possibility that such a totipotent cell might be implanted into a woman's uterus and be used for human reproductive cloning, to "make a copy" of a living person. In the opinion of virtually everyone, scientists and nonscientists alike, this would be highly unethical. For starters, it would be unsafe, as virtually all other mammalian clones (including Dolly) have turned out to be abnormal in some way. (Dolly suffered from arthritis and obesity, and died early.) Moreover, it raises serious concerns about the burden of expectations placed upon the new "clone." It does not address any compelling need that can be readily defended on moral grounds. Finally, to me and many others, it crosses a line that is deeply troubling: human reproduction should not deviate from the union of sperm and egg.

But would the production of a human totipotent cell line by SCNT be itself unethical? After all, it is hard to attach moral significance to a skin cell, and it is also hard to see how a human egg cell with its nucleus removed, now simply a bag of cytoplasm, has moral standing. So how does the fusion of those two entities in the laboratory, a very unnatural event, acquire such status? If uterine implantation of such a cell line was absolutely prohibited, then many thoughtful observers, including religious believers like me, could defend human SCNT research.

More recently, however, an even more dramatic development in stem cell research has appeared that may provide powerful solutions to both the practical and the ethical issues. Working in Japan, Dr. Shinya Yamanaka reasoned from the experience with SCNT that the cytoplasm of an egg must provide a finite set of signals capable of re-programming a human skin cell. If that set of signals could be identi-fied, then perhaps stem cells could be created much more simply, with no need to utilize egg donors. Most other investigators thought that this was a pipe dream, but Yamanaka persisted.

Only a few times in my career have I witnessed a scientific devel-opment that qualifies as a true revolution. But that was my reaction when I read Yamanaka's 2006 paper, showing that the transfer of just four genes was capable of transforming a mouse skin cell into a pluri-potent stem cell, capable of differentiating into virtually all mouse tissues. Just a year later, Yamanaka and another group showed that the same could be done with human skin cells; and more recently it has been possible to show that this can be achieved using cells from a single human hair!

This new cell type, referred to as an induced pluripotent stem (iPS) cell (Figure 10.3C), has opened a new window for research and potential clinical applications. In the short time since Yamanaka's protocol was published, others have managed to reduce the number of genes required for this transformational process down to as few as one, and have rapidly generated such iPS cell lines from skin cells of individuals with a long list of inherited diseases. Those cells now provide us with a golden opportunity to understand in detail how those diseases cause the damage they do, without having to put living humans at risk.

The therapeutic use of iPS cells is still very uncertain, but holds great promise. After all, iPS cells can be derived from anyone, and so could be transplanted back into the same person with no risk of rejection. This brings self-healing to a whole new level.

Before we get too carried away about the potential of iPS cells to cure a long list of degenerative diseases, however, it should be mentioned that there are serious existing concerns about whether such cells might be, shall we say, a bit too exuberant in their potential, actually causing tumors in the person receiving them. That risk is heightened by the fact that one of the genes used to achieve the pluripotent state is an oncogene.

It will be several years before the potential of iPS cells for therapeutics can be determined, but it is entirely possible that this approach will find clinical applications for a variety of illnesses in which tissues are dying and replacement is needed. The conditions could include disorders of the liver, heart, kidneys, islet cells (which make insulin) in the pancreas, and even the brain.

There could even be protocols that combine gene therapy with iPS cells. For instance, for the treatment of sickle-cell anemia, a skin cell could be taken from a patient, transformed into an iPS cell, and then "fixed" with *ex vivo* gene therapy, eliminating the sickle mutation. Those cells could then be treated in the laboratory to differentiate them into bone marrow stem cells, and then infused into the original patient (perhaps after treatment with chemotherapy to "make space" for the transfused bone marrow). This protocol may sound like science fiction, but it has already been successfully carried out in a mouse with sickle-cell anemia. The mouse was cured.

ENABLING THE FUTURE

These two therapeutic developments—gene therapy and stem cell therapy—hold great but uncertain promise for the future. I am willing to make some bets on them, however. Combining their promise with all the other components of personalized medicine in diagnostics, prevention, and therapeutics, as described in the preceding chapters, portends a dramatic change in medicine. One should keep in mind the "first law of technology": the consequences of a radical new technology are almost always overestimated in the short term and underestimated in the long term. Still, let's follow the course of a hypothetical member of the human race in the twenty-first century.

She was born on January 1, 2000, a child of the new millennium. Her parents named the little girl Hope.

Hope had a typical small-town childhood, graduated from a public high school, and went on to college. Like most kids, Hope didn't think much about mortality, but that perspective was shattered when she was 20 and her favorite uncle died, at age 48, of a heart attack. Distressed by this loss, and concerned for the rest of the family, Hope's mother initiated a detailed investigation of the medical history of the extended family. She used an upgraded version of the original Surgeon General's family medical history tool to collect all this information into an electronic form, which was then shared with the rest of the family. The survey documented that Hope's maternal grandfather had also died of a heart attack. There were additional individuals on her father's side with diabetes and cancer.

Encouraged by her mother, Hope made an appointment with a primary care physician to discuss what steps she might take to address her potential heart disease risk.

Hope's doctor studied the family medical history, and agreed that this was an opportune time to do a full evaluation. In addition to standard blood tests, he proposed a complete genome sequencing analysis, which by the year 2020 cost only $300. Hope's parents and younger brother also had such analysis done.

Hope received her results electronically a month later. A bit over-whelmed by the amount of detail, she returned to talk with a physi-cian assistant (PA). The PA had specialized in personalized medicine, and handled such consultations frequently. Among other things, Hope learned that she was a carrier for cystic fibrosis, that relative to the average person she had a slightly increased risk of breast cancer, and that she carried genetic factors placing her at modestly increased risk for high blood pressure, especially if she became overweight.

The most dramatic finding, however, was a series of genetic varia-tions that predicted her heart attack risk as threefold higher than the average, despite the fact that her cholesterol levels were normal. In her case, apparently it was not the lipids in her blood that represented the problem. Instead, her blood platelets, normally responsible for blood clotting, were too "sticky." This stickiness might have been an ad-vantage for her Stone Age ancestors who faced daily risks of major trauma, but for Hope it predisposed her to future blockages in her coronary arteries.

Hope and the PA discussed ways to prevent a future heart attack by a program of diet, exercise, and the use of a platelet inhibitor drug.

Though Hope had previously not been very motivated to pay at-tention to health maintenance, the news about her heart disease risk was a kick in the pants. Her new motivation was enhanced by learn-ing that if she documented her attention to diet and exercise, her medical insurance copayments would be reduced.

Five years later, Hope met Mr. Right. George was impressed by her daily regimen, and had to admit he hadn't paid much attention to risks to his own health. But after their engagement, he decided it was time for him to think more about his own future, and so he also compiled a family history. He learned that family members had been affected with diabetes, obesity, Parkinson's disease, and colon cancer. He too decided to proceed to genome sequencing.

Hope held her breath, worried that George would also be a CF carrier and that this news would force some difficult decisions about childbearing. When the report came back, George's CF gene was normal. But other aspects of his DNA report included a twofold increased risk for obesity and colon cancer. Hope also learned that George carried a variant that had been associated with reduced marital fidelity—but she convinced herself that this was a statistical finding of no real significance for her obviously smitten fiancé. The wedding went on as planned.

Three years later, Hope and George decided to start a family. They knew they did not have to fear cystic fibrosis, or any other devastating single-mutation disease. Nonetheless, they considered the possibility of preimplantation genetic diagnosis (PGD) to screen potential embryos for some of the risky variants that they would just as soon not pass on. Yet Hope and George decided that none of the risks were high enough to justify such interventions, and it was a lot more fun to make babies the old-fashioned way. Their new son, Raymond, whom George delighted in referring to as his little "Ray of Hope," was apparently healthy in every way. Hope and George were asked to give their permission for a genome sequence to be conducted on their newborn son, and they readily agreed. They were told that they would be informed only about conclusions that required potential action by the parents during childhood. The rest of the information would be gradually revealed as the scientific basis for its conclusions was established.

The report on baby Ray's genome was generally reassuring. But there was a 60 percent risk of obesity, as Ray had inherited George's predispositions, together with additional risk variants from Hope. Working with the pediatrician, Hope and George designed a diet for Ray that was reduced in fat and calories compared with the standard nutritional recommendations for newborns, but still adequate to support normal growth and development. As interactions between individuals and the microbes in their gastrointestinal tract had also been shown to play a role in obesity, Ray's diet was supplemented with a probiotic formula that recolonized his intestine with a mix of microorganisms to help promote normal body weight.

By 2035, all three members of this little nuclear family were doing well. Like many people at that time, they had each started wearing "smart shirts," which carried embedded sensors and transmitters to keep track of various body variables, connected wirelessly to their health care providers to alert these providers of any troublesome findings (Figure 10.4). The smart shirt also enabled Hope to keep track of Ray's exercise and dietary intake, successfully keeping him in a healthy weight range.

Figure 10.4: The home health station of the future, equipped with a monitor for recording body parameters; a saliva cup to monitor body metabolites and the oral microbiome; an exercise station that allows imaging of the person at rest and during exercise; and an environmental sampler for the indoor and outdoor air.

In 2045, George had his first noninvasive exam for colon polyps, given his genetic risk of colon cancer. Sure enough, two polyps were discovered and removed. If these had remained undiscovered for a few more years, they might have gone on to develop into cancer.

As the years passed, the potential for extending the human life span grew. Hope and George began to explore the possibility of taking a new drug that had just been approved for that purpose.

In 2055, Hope's mother died. She and Hope had always been very close. Hope knew this death was coming, but nevertheless found it very hard to adjust to her loss. After a few months of tearfulness and sadness, she sought help from her care provider, who recognized that she was experiencing a prolonged situational depression. Referring to her genome sequence, the provider concluded that she was particularly susceptible to this kind of outcome. She also determined that one drug in particular would be most suitable and could be optimized according to Hope's specific metabolism. Within two weeks Hope realized that the fog was lifting. After a few more weeks, she was able to stop taking the drug, proceed with normal grieving, and return to her usual self.

Through it all, Hope continued her dietary and exercise regimen, aware that her heightened risk of heart disease lurked in the background. (George was not so disciplined—he exercised too little, and drank a bit too much. Yet his health remained good, as did his marital fidelity.) When Hope was age 68, it happened. While gardening, she experienced a sudden onset of pain in her left arm, sweating, nausea, and shortness of breath. Moments later, emergency medical technicians arrived at her front doorstep, having been alerted by her smart shirt of an impending heart attack. With her genome sequence as a guide, they immediately instituted the proper drug treatment, heading off what otherwise might have been a fatal event.

The next year, George's own risks caught up with him. He developed the early signs of Parkinson's disease. Now, at age 70, he feared the worst. But he worried unnecessarily. Doctors transformed George's own skin cells into just the kind of neuronal cells needed in his brain to reverse the disease. With the aid of nanotechnology robots that were programmed to deliver the cells to exactly the right location, George's symptoms were readily corrected. Though he needed another treatment two years later, his neurologist was unable to detect any signs of Parkinson's disease when George came in for his annual physical at age 90.

On January 1, 2100, Hope celebrated her hundredth birthday. Later that year their family gathered around Hope and George for their seventy-fifth wedding anniversary, and wished them well for many more years together.

COULD THE DREAM BECOME A NIGHTMARE?

The story of Hope, George, and Ray is, of course, completely fictional, and the details become more and more imaginative as we travel past 2025. The lifesaving components of this story are exhilarating to contemplate, but they are not guaranteed to come true. Imagine, if you will, a very different biography of Hope . . .

When Hope's uncle died, no educational information was available to help the family members explore their own risks. When Hope asked her provider about the possibility of practicing better preventive medicine, he warned her that such interventions would not be reimbursed, and that a lack of effective research studies meant that the evidence for utility was limited. Hope had heard that genome sequencing was now quite inexpensive, but her doctor thought it was just overblown marketing and recommended against it. Hope met George, they married, and they had a son. It was clear that Ray was becoming seriously obese by age six. A lifelong struggle to change that pattern ensued, without much success.

Never having been given a chance to develop her own plan for staying healthy, Hope avoided exercise, ate an unhealthy diet, and gained weight. By age 35 she had developed high blood pressure. Her physician, uninformed about the possibility of individualized drug therapy, prescribed a drug that was wrong for her and caused an adverse reaction. She concluded that physicians don't know what they're doing, and stopped the therapy.

At age 50, working in her garden on a hot day, she experienced an onset of pain in the left arm. No smart shirts had been developed. George called her doctor, who consulted a standard "one size fits all" algorithm, and said that Hope was too young to be having cardiac symptoms. It must be a pulled muscle, he said. Hope arrived in the emergency room two hours later, in shock. A major section of her heart muscle had already died. Although stem cell therapy might potentially have been helpful in repairing the damage, there was no time to generate such cells from her. Despite the best efforts of the emergency room staff, Hope suffered a cardiac arrest and could not be resuscitated. She had lived less than half of her potential life span. Grieving at her bedside were Ray, now morbidly obese, and George, unaware that his undiagnosed colon cancer was about to spread to his liver.

What a grim scenario! Sadly for us all, this disappointing outcome could still happen. Yes, medical science, built upon ever-increasing knowledge of the human genome, is poised to deliver substantial medical benefits in the coming years. Good science is necessary but not sufficient—it will take the full engagement of researchers, governments, health care providers, and the general public to avoid this depressing alternative.

PLANNING FOR SUCCESS

Our shared motto should be: *Keep Hope Alive!* Because, you see, Hope is you. And Hope is me. And Hope is our partners, our siblings, our children, our grandchildren, our nieces and nephews, and our friends. The stakes are nothing less than our shared dream of living life to the fullest.

So what do we need to do to keep Hope alive?

1. **Research.** We live in an exciting time in medical research, where a series of "disruptive innovations" promise to revolutionize diagnosis, prevention, and treatment. But this revolution will sputter out if we do not make continued major investments in the research needed to explore these new approaches. Worldwide, medical research is seriously underfunded in nearly all countries. Most companies will tell you that at least 15 percent of their revenue should be invested in research and development. Yet in the United States, spending on medical research adds up to only about 5 percent of the cost of health care. Despite this stingy attitude, the benefits of research are clear. Life expectancy has increased 6 years over the course of the last 30 years. Deaths from heart disease have fallen 63 percent in the last 30 years, with an investment in heart disease research of just $3.70 per American per year. Polls have repeatedly shown that the American public believes in medical research and is willing to see more tax dollars invested in it. Yet at the time I write this, fewer than one grant application out of five submitted to the National Institutes of Health will be funded. In consequence, many young medical researchers are increasingly demoralized and likely to leave the field. We must do better.

2. **The electronic medical record.** Given the need to correlate increasingly large data sets of DNA sequence, medical data, and environmental exposures, it is inconceivable that the full power of

personalized medicine can be achieved without an electronic medi-cal record. It is shameful that we have come as far as we have with-out tackling this problem. Can you imagine going to your bank to discover that it keeps all its financial records on random pieces of paper? And yet in many physicians' offices, that is the way this precious information is recorded. The consequences are almost entirely negative. Getting copies of your own medical records can take weeks or months, and often costs a ridiculous sum of money. Obtaining your own medical information if you happen to need it in an emergency in another town or another state can be almost impossible. Private enterprises such as Google Health and Microsoft HealthVault are stepping in, giving individuals an opportunity to store electronic copies of their own medical records, with the abil-ity to control who sees them and when. But the full solution to this problem will not be achieved until all medical records are produced in electronic form, with appropriate safeguards to prevent breaches of privacy. Sure, this is complicated, but it's a solvable problem! If we can do this for credit card records and bank records, why on earth can we not do it for medical records?

3. **Good policy decisions.** Science moves forward at an ever-increasing pace, but the process for establishing appropriate policies for implementation of new scientific findings can be frustratingly slow. A recent survey of the most frequently cited papers in the lit-erature about medical interventions revealed that the median time lag between the earliest publication of a medical discovery and its implementation was 24 years! This horrendous track record must be improved. Government overseers are charged with evaluating ef-fectiveness of new interventions, and that cannot be left solely to the marketplace—but oversight must continually strive to maintain a balance between protecting the public and encouraging innovation. All too often, that process has tilted toward the conservative side,

depriving the public of promising innovations for many years, until a very large number of expensive and duplicative studies are carried out. Our health care system must also be prepared to shift from a focus on treating advanced disease toward reimbursing for preventive care, especially when that can be economically documented to show benefit. And personalized medicine should welcome a bright light of oversight shining on its value proposition.

4. **Education.** Personalized medicine depends intimately upon the principle that knowledge is power, both for you and for your doctor. And yet the science of genomics has come along very rapidly, and many doctors are unprepared to take full advantage of it. From reading this book, you almost certainly know more than your doctor about personalized medicine. Most physicians, physician assistants, nurse practitioners, nurses, and other health care providers have very limited understanding of these new principles, and a major effort is needed to bring them quickly up to speed. The National Coalition for Health Professional Education in Genetics (NCHPEG), an organization that I helped cofound along with the American Medical Association and the American Nurses Association, is trying to fill that void, but has very limited resources and continually faces the challenge of getting attention from busy practitioners who think genetics is irrelevant to their practice.

5. **Ethical decision making.** Alleviating human suffering is an ethical mandate shared by nearly all cultures and faiths throughout history. Any argument about the need to slow down medical research must therefore be considered in the context of depriving suffering individuals of some potential benefit. Yet there are certain research applications, such as human reproductive cloning, that are considered unethical by virtually all who have looked at the circumstances closely. There are many other areas, however, where it is difficult to draw a bright line. In our pluralistic society, it is not always easy to

arrive at an ethical consensus. Over the last decade and a half the Presidential Bioethics Commission in the United States has provided a venue for healthy debate about some of these issues, but the limited authority of the group and the strong influence of politics in appointing its members have diminished its impact. A more authoritative model, such as the one currently functioning in the United Kingdom, the Human Fertilization and Embryology Authority, should be strongly considered in the future.

A FINAL EXHORTATION

Remember Saint-Exupéry: "As for the future, your task is not to foresee, but to enable it." For the future of personalized medicine, this exhortation is not just for the scientific community, or the medical community, or the government—it is for each of us. The success of personalized medicine will come about only when we each take responsibility for our health. Health care providers can help, but they cannot drive your bus. Each chapter in this book, including this one, has concluded with a list of things you can do now to take full advantage of the potential for personal empowerment. If you follow those recommendations, you will truly be on the leading edge of this new revolution. But the edge will keep moving, and so it will be essential to upgrade your own knowledge base periodically.

Wayne Gretsky is considered by many to be the greatest hockey player of all time. His father, Walter, was his first and best coach, and his most sage advice was simple: "Skate where the puck is going to be." We are all skaters in the game of life. We're trying to move effectively on the ice, work with our team, avoid falling, and even score a few goals. But it's not enough to set your sights on following the puck around. You need to skate where the puck is going to be. Your DNA helix, your language of life, can also be your own textbook of medicine. Learn to read it. Learn to celebrate it. It could save your life.

WHAT YOU CAN DO NOW TO JOIN THE PERSONALIZED MEDICINE REVOLUTION

Collecting the medical records of yourself and your dependents, and converting them to electronic form so that they can be rapidly accessed by care providers that you designate, can save all sorts of headaches, and can even be lifesaving in an emergency. Two Web-based providers of this service that you might want to consider are Google Health (www .google.com/health) and Microsoft HealthVault (www.healthvault.com).

Now that you have reached the conclusion of the *Language of Life*, go back and look at "What You Can Do" in each of the previous chapters. Perhaps there were suggestions that didn't appeal to you at the time—but they might now. Take advantage of your new knowledge to do everything you can to focus on staying healthy.

Hollywood's view of advances in genetics may be different from what the facts would warrant. Just for fun, rent the movie *GATTACA* and make a list of all the aspects of the film that are scientifically misguided.

Congratulations—you are now in the top 1 percent of individuals in terms of your understanding of personalized medicine. So use this information to start conversations with family and friends. Spread the word!

Glossary

(Excerpted fom the public "Talking Glossary of Genetic Tems,"
available from the National Human Genome Research Institute at
http://genome.gov/glossary.cfm.)

ACGT: Refers to the four types of bases in a DNA molecule. The letters are abbreviations for the chemical names: adenine, cytosine, guanine, and thymine. A DNA molecule consists of two strands wound around each other. The strands are held together by bonds between the bases. A pairs with T, and C pairs with G. The sequence of bases in a portion of a DNA molecule, called a gene, carries the instructions needed to assemble a protein.

Allele: One of two or more versions of a gene. An individual inherits two alleles for each gene: one allele from each parent. If the two alleles are the same, the individual is said to be homozygous for that gene. If the alleles are different, then the individual is heterozygous for that gene. Although the term "allele" was originally used to describe variation among genes, today it also can refer to variation among non-coding DNA sequences.

Amino Acids: A set of 20 different small molecules used to build proteins. Proteins consist of one or more chains of amino acids called polypeptides. The sequence of the amino acid chain causes the polypeptide to fold into a shape that is biologically active. The amino acid sequences of proteins are encoded in the genes.

Base Pair: Two chemical bases bonded to each other, forming a "rung" of the DNA "ladder." The DNA molecule consists of two strands that wind around each other like a twisted ladder. Each strand has a backbone made of alternating sugar (deoxyribose) and phosphate groups. Attached to each sugar is one of four bases A, T, C, or G. The two strands are held together by bonds between the bases (A forms a base pair with T, and C forms a base pair with G).

BRCA1/BRCA2: The first two genes found to be associated with inherited forms of breast cancer. In healthy people, both genes play roles as tumor suppressors—that is, they help regulate cell division. When these genes are rendered inactive owing to mutation, uncontrolled cell growth results, leading to the cancer. Women with mutations in either gene have a much higher risk for developing breast and ovarian cancer than women with normal versions of the genes.

Carrier: An individual who carries and may pass on a genetic mutation associated with a disease but does not display symptoms of that disease. Carriers are associated with diseases inherited as recessive traits. In order to have the disease, an individual must have inherited mutated alleles from both parents. An individual having one normal allele and one mutated allele is a carrier and does not have the disease. Two carriers may produce children with the disease.

Carrier Screening: A type of genetic screening performed on people who display no symptoms of a recessive genetic disorder but may be at risk for passing it on to their children. A carrier for a genetic disorder has inherited one normal and one abnormal allele for a gene associated with the disorder. A child must inherit two abnormal alleles in order for symptoms to appear.

Chromosome: An organized package of DNA found in the nucleus of the cell. Different organisms have different numbers of chromosomes. Humans have 23 pairs of chromosomes: 22 pairs of numbered chromosomes called autosomes, and one pair of sex chromosomes (X and Y). Each parent contributes one chromosome to each pair so that offspring get half of their chromosomes from the mother and half from the father.

Cloning: A process for making identical copies of an organism, cell, or DNA sequence. Molecular cloning is a process by which scientists amplify a desired DNA sequence. The sequence of interest is isolated, inserted into another DNA molecule, called a vector, and introduced into a suitable host cell. Each time the host cell divides, it replicates the foreign DNA sequence along with its own DNA. Cloning also can refer to asexual reproduction.

Cytoplasm: The gelatinous liquid that fills the inside of a cell. It is composed of water, salts, and various organic molecules. Some intracellular organelles, such as the nucleus and mitochondria, are enclosed by membranes that separate them from the cytoplasm.

Deletion: A type of mutation involving the loss of genetic material. A deletion mutation can be small, involving a single missing DNA base pair, or large, involving a piece of a chromosome.

Diploid: See Haploid.

DNA (deoxyribonucleic acid): The chemical name for the molecule that carries genetic instructions in all living things. The DNA molecule consists of two strands that wind around each other in a double helix. Each strand has a backbone made of alternating sugar (deoxyribose) and phosphate groups. Attached to each sugar is one of four bases: A, T, C, or G. The two strands are held together by bonds between the bases (A bonds with T, and C bonds with G). The sequence of the bases along the backbones serves as instructions for assembling protein and RNA molecules.

DNA sequencing: A laboratory technique for determining the exact sequence of bases (A, T, C, and G) in a DNA molecule. The DNA base sequence carries the information the cell needs to assemble RNA and protein molecules. Therefore, information about DNA sequence is important to scientists investigating the functions of genes. The technology of DNA sequencing was made faster and less expensive as a part of the Human Genome Project.

Enzyme: A biological catalyst. An enzyme is almost always a protein. It speeds up the rate of a specific chemical reaction in the cell. The enzyme is not destroyed during the reaction and is used over and over. A cell contains thousands of different types of enzyme molecules, each specific for a different chemical reaction.

Exon: A portion of a gene that codes for amino acids. In the cells of plants and animals, most gene sequences are broken up by one or more DNA sequences called introns. The parts of the gene sequence that are expressed in the protein are called exons (because they are expressed), whereas the parts of the gene sequence that are not expressed in the protein are called introns (because they come in between the exons).

Founder effect: Refers to the reduction in genetic variation that results when a small subset of a large population is used to establish a new colony. The new population may be different from the original population, in terms of both its genotypes and its phenotypes.

Frameshift mutation: A type of mutation involving the insertion or deletion of a DNA sequence, where the number of base pairs is not divisible by three. "Divisible by three" is important because the cell reads a gene in groups of three bases. Each group of three bases corresponds to one of the 20 different amino acids used to build a protein. If a mutation disrupts this reading frame, then the entire DNA sequence following the mutation will be read incorrectly.

Fraternal twins: see Identical twins.

Gene: The basic physical unit of inheritance. Genes are passed from parents to offspring and contain the information needed to specify traits. Genes are arranged, one after another, on structures called chromosomes. A chromosome contains a single long DNA molecule, only a portion of which corresponds to a single gene. Humans have approximately 20,000 protein-coding genes arranged on their chromosomes.

Gene mapping: The process of establishing the locations of genes on the chromosomes. Early gene maps used linkage analysis. The closer two genes are to each other on the chromosome, the more likely it is that they will be inherited together. By following inheritance patterns, the relative positions of genes can be determined. More recently, scientists have used recombinant-DNA techniques to establish the actual physical locations of genes on the chromosomes.

Gene therapy: An experimental technique for treating disease by altering the patient's genetic material. Most often, gene therapy works by introducing a healthy copy of a defective gene into the patient's cells.

Genetic drift: A mechanism of evolution. It refers to random fluctuations in the frequencies of alleles from generation to generation due to chance events. Genetic drift can cause traits to either become prominent in or to disappear from a population. The effects of genetic drift are most pronounced in small populations.

Genetic engineering: The process of using recombinant-DNA technology to alter the genetic makeup of an organism. Traditionally, humans have manipulated genomes indirectly by controlling breeding and selecting offspring with desired traits. Genetic engineering involves the direct manipulation of one or more genes. Most often, a gene from another species is added to an organism's genome to give it a desired phenotype.

Genetic marker: A DNA sequence with a known physical location on a chromosome. Genetic markers can help link an inherited disease with the responsible gene. Segments of DNA close to each other on a chromosome tend to be inherited together. Genetic markers are used to track the inheritance of a nearby gene that has not yet been identified, but whose approximate location is known. The genetic marker itself may be a part of a gene or may have no known function.

Genetic screening: The process of testing a population for a genetic disease in order to identify a subgroup of people who either have the disease or have the potential to pass it on to their offspring.

Genetic testing: The use of a laboratory test to look for genetic variations associated with a disease. The results of a genetic test can be used to confirm or rule out a suspected genetic disease or to determine the likelihood that a person will pass on a mutation to his or her offspring. Genetic testing may be performed prenatally or after birth. Ideally, a person who undergoes a genetic test discusses the meaning of the test and its results with a genetic counselor.

Genome: The entire set of genetic instructions found in a cell. In humans, the genome consists of 23 pairs of chromosomes found in the nucleus as well as a small chromosome found in the cells' mitochondria. These chromosomes, taken together, contain approximately 3.1 billion bases of DNA sequence.

Haploid: Refers to a cell or an organism having a single set of chromosomes. Organisms that reproduce asexually are said to be haploid. Sexually reproducing organisms are said to be diploid (having two sets of chromosomes: one from each parent). Only their egg and sperm cells are haploid.

Haplotype: A set of DNA variations (or polymorphisms) that tend to be inherited together. A haplotype can be a combination of alleles or a set of single nucleotide polymorphisms (SNPs) found on the same chromosome. Information about haplotypes is collected by the International HapMap Project and is used to investigate the influence of genes on disease.

HapMap: An international project that seeks to relate variations in human DNA sequences with genes associated with health. A haplotype is a set of DNA variations (or polymorphisms) that tend to be inherited together. A haplotype can be a combination of alleles or a set of single nucleotide polymorphisms (SNPs) found on the same chromosome. The HapMap describes common patterns of genetic variation among people.

Identical twins: Also called monozygotic twins. They result from the fertilization of a single egg that splits very soon afterward. Identical twins share all their genes and are always of the same sex. In contrast, fraternal twins result from the fertilization of two separate eggs during the same pregnancy. They share half of their genes, just like any other siblings. Fraternal twins may be of the same or different sexes.

Intron: A portion of a gene that does not code for amino acids. In the cells of plants and animals, most gene sequences are broken up by one or more introns. The parts of the gene sequence that are expressed in the protein are called exons (because they are expressed), whereas the parts of the gene sequence that are not expressed in the protein are called introns (because they come in between the exons).

Karyotype: An individual's collection of chromosomes. The term also refers to a laboratory technique that produces a photograph of an individual's chromosomes. The karyotype is used to look for abnormal numbers or structures of chromosomes.

Messenger RNA (mRNA): A single-strand RNA molecule that is complementary to one of the DNA strands of a gene. The mRNA is an RNA version of the gene that leaves the cell nucleus and moves to the cytoplasm where proteins are made. During protein synthesis, an organelle called a ribosome moves along the mRNA, reads its base sequence, and uses the genetic code to translate each 3-base triplet into its corresponding amino acid.

Mutation: A change in a DNA sequence. Mutations can result from DNA copying mistakes made during cell division, from exposure to ionizing radiation, exposure to chemicals called mutagens, or infection by viruses. Germ line mutations occur in the eggs and sperm and can be passed on to offspring, whereas somatic mutations occur in body cells and are not passed on.

Non-coding DNA: DNA sequences that do not code for amino acids. Most non-coding DNA lies between genes on the chromosome and is of unknown function. Other non-coding DNA, called introns, is found within genes. Some non-coding DNA plays a role in the regulation of gene expression.

Nucleic acid: An important class of macromolecules found in all cells and viruses. The functions of nucleic acids have to do with the storage and expression of genetic information. Deoxyribonucleic acid (DNA) encodes the information the cell needs to make proteins. A related type of nucleic acid, ribonucleic acid (RNA), carries information to the cytoplasm to participate in protein synthesis.

Oncogene: A mutated gene that contributes to the development of a cancer. In their normal, unmutated state, onocgenes are called proto-oncogenes, and they play roles in the regulation of cell division. Oncogenes work like putting your foot down on the accelerator of a car, pushing a cell to divide.

Pharmacogenomics: A branch of pharmacology that is concerned with using data about DNA and amino acid sequence to inform drug development and testing. An important application of pharmacogenomics is correlating individual genetic variation with drug responses.

Polygenic trait: A trait whose phenotype is influenced by more than one gene. Traits that display a continuous distribution, such as height or skin color, are polygenic. The inheritance of polygenic traits does not show the phenotypic ratios characteristic of Mendelian inheritance, although each of the genes contributing to the trait is inherited as described by Mendel. Many polygenic traits are also influenced by the environment and are called multifactorial.

Protein: An important class of molecules found in all living cells. A protein is composed of one or more long chains of amino acids; the sequence is a translation of the DNA sequence of the gene that encodes that protein. Proteins play a variety of roles in the cell, including structural (cytoskeleton), mechanical (muscle), biochemical (enzymes), and cell signaling (hormones). Proteins are also an essential part of the diet.

Race: In common parlance, a group of people who share a set of visible characteristics such as skin color, facial features, and hair texture, as well as a sense of identity. Although these visible traits are influenced by genes, the vast majority of genetic variation exists within racial groups and not between them. For this reason, many scientists believe that race is more accurately described as a social construct, not a biological one.

Recessive: Refers to a relationship between two versions of a gene. Individuals receive one version of a gene, called an allele, from each parent. In the case of a recessive genetic disorder, an individual must inherit two copies of a mutated allele in order for the disease to be present.

Recombinant DNA: A technology that uses enzymes to cut and paste together DNA sequences of interest. The recombined DNA sequence can be placed into vehicles called vectors that ferry the DNA into a suitable host cell, where it can be copied or expressed.

Retrovirus: A type of virus that uses RNA as its genetic material. When a retrovirus infects a cell, it makes a DNA copy of its genome that is inserted into the DNA of the host cell. There are various different retroviruses that cause human diseases, including AIDS.

RNA (ribonucleic acid): A molecule similar to DNA. Unlike DNA, RNA is a single strand. An RNA strand has a backbone made of alternating sugar (ribose) and phosphate groups. Attached to each sugar is one of four bases: A, U, C, or G. Different types of RNA exist in the cell: messenger RNA, ribosomal RNA, and transfer RNA. More recently, some small RNAs have been found to be involved in regulating gene expression.

Stem cell: A type of cell with the potential to form many of the different cell types found in the body. When stem cells divide, they can form more stem cells, or other cells that perform specialized functions. Embryonic stem cells are pluripotent and have the potential to form a complete individual, whereas adult stem cells are multipotent and can form only certain types of specialized cells. Stem cells continue to divide as long as the individual remains alive.

Telomere: An end of a chromosome. Telomeres are made of repetitive sequences of non-coding DNA that protect the chromosome from damage. Each time a cell divides, the telomeres become shorter unless a repair enzyme called telomerase is present. Eventually, the telomeres become so short that the cell can no longer divide.

Transgenic: Having one or more DNA sequences from another species, introduced by artificial means. Animals usually are made transgenic by having a small sequence of foreign DNA injected into a fertilized egg or developing embryo. Transgenic plants can be made by introducing foreign DNA into a variety of different tissues.

Tumor suppressor gene: A gene whose normal function is to direct the production of a protein that is part of the system that slow down cell division. The tumor suppressor protein plays a role in keeping cell division in check. When mutated, a tumor suppressor gene is unable to do its job. As a result, uncontrolled cell growth may occur, contributing to the development of a cancer.

Variant: A spelling difference in DNA.

Vector: Any vehicle (often a virus or a plasmid) that is used to ferry a desired DNA sequence into a host cell as part of a molecular cloning procedure. Depending on the purpose of the cloning procedure, the vector may assist in multiplying, isolating, or expressing the foreign DNA insert.

Virus: An infectious agent that occupies a place near the boundary between the living and the nonliving. A virus is a particle much smaller than a bacterial cell. It consists of a small genome of either DNA or RNA surrounded by a protein coat. Viruses enter host cells and hijack the enzymes and materials of the host cells to make more copies of themselves. Viruses cause a wide variety of diseases in plants and animals, including AIDS.

X chromosome: One of two sex chromosomes. Humans and most mammals have two sex chromosomes: the X and the Y. Females have two X chromosomes in their cells; males have an X and a Y chromosome in their cells. Egg cells all contain an X chromosome; sperm cells may contain an X or a Y chromosome. This arrangement means that during fertilization, it is the male that determines the sex of the offspring.

X-linked, or sex-linked: Term used for a trait whose gene is located on the X chromosome. Humans, and most mammals, have two sex chromosomes: the X and the Y. In a sex-linked disease, it is usually males that are affected because they have a single copy of the X chromosome that carries the mutation. In females, the effect of the mutation may be masked by the second, healthy copy of the X chromosome.

Y chromosome: One of two sex chromosomes. Humans and most mammals have two sex chromosomes: the X and the Y. Females have two X chromosomes in their cells; males have an X and a Y chromosome in their cells. Egg cells all contain an X chromosome; sperm cells may contain an X or a Y chromosome. This arrangement means that during fertilization, it is the male that determines the sex of the offspring.

Genetics 101

Descriptions of the basic principles of genetics, genomics, and molecular biology in Chapter 1 were limited to the fundamentals necessary for an understanding of how DNA plays a role in health and disease. Appendix B provides a bit more information for the interested reader who wants to go a little deeper.

DNA, the Language of Life

As was shown in Figure 1.1, DNA is a long organic polymer. The double strand of DNA provides elegant redundancy for the information; A on one strand is always paired with T on the other, and likewise G and C are always partnered, held together by weak chemical forces between these bases that prevent the entire double helix from falling apart. Among other benefits, this redundancy provides protection against damage; if a base is zapped by a passing cosmic ray, for instance, the information on the other strand provides an immediate template for repairing the damage. Furthermore, as Watson and Crick pointed out, and as was subsequently demonstrated experimentally, this double-strand structure provides an elegant mechanism for copying DNA. Unwinding the double helix and using each strand as a template for the synthesis of a new one (a process carried out with high fidelity and great speed by the enzyme DNA polymerase) provide an ideal copying mechanism each time the cell divides.

If DNA is the instruction book, how are the instructions carried out? Think of DNA as an encyclopedia. Printed out, the information inside each human cell would fill about 400 volumes, about 20 times more than the full *Encyclopaedia Britannica*. But just as the *Encyclopaedia Britannica* is divided up into entries, so the genome contains its own packets of information, called genes. In its simplest form, a gene is an instruction for a particular function. The function of the best-known genes is carried out by having that DNA transcribed into RNA, and subsequently translated into protein (shown in Figure 1.2).

This "central dogma of molecular biology" depicts RNA as a messenger that carries the information encoded within DNA out of the nucleus of the cell and into the cytoplasm. There it encounters an amazing protein factory, the ribosome. You can think of the ribosome as an elegant molecular decoder, using a translation dictionary that is essentially consistent across all species, translating every three RNA bases into a protein subunit called an amino acid. With $4 \times 4 \times 4 = 64$ possible triplet "codons" in the RNA, and only 20 amino acids in the protein, room is left for redundancy. And that's just what is observed. For instance, AAA and AAG in RNA both code for the amino acid lysine, and AGA and AGG both code for arginine.

Now that the complete sequence of the human genome has been determined, it is possible to count up the number of genes that code for proteins in this fashion. Before we actually had the instruction book in front of us, those estimates varied widely, with the average guess by molecular biologists being about 100,000. In fact, genome scientists organized a sweepstakes in 1999, as a fun-loving way to try to encourage scientists to stick their necks out. The range of guesses extended from 30,000 to 150,000; my own guess was 48,004. (Okay, I was trying to avoid a round number.) Imagine the surprise and consternation of the scientific community when the ultimate answer was a mere 20,000 protein-coding genes. Having gotten used to the idea that total genome size was not predictive of organismal complexity, many scientists still expected that at least the gene count would reflect this. But those expectations were not realized; after all, the lowly roundworm, a subject of much research interest because it serves as a model organism for understanding developmental biology, has a tally of about 19,000 genes, and the current count of genes for rice is substantially higher than that for humans. So if you were relying on the gene count to validate the superiority of *Homo sapiens*, think again. There are lots of things on your dinner plate this evening that have more genes than you do.

The analogy of the *Encyclopaedia Britannica* begins to break down when you look more closely at the organization of the genome. It turns out that only about 1.5 percent of the human genome is involved in coding for protein. But that doesn't mean the rest is "junk DNA." A number of exciting new discoveries about the human genome should remind us not to become complacent in our understanding of this marvelous instruction book. For instance, it has recently become clear that there is a whole family of RNA molecules that do not code for protein. These so-called non-coding RNAs are capable of carrying out a host of important functions, including modifying the efficiency by which other RNAs are translated. In addition, our understanding of how genes are regulated is undergoing dramatic revision, as the signals embedded in the DNA molecule and the proteins that bind to them are rapidly being elucidated (Figure B.1). The complexity of this network of regulatory information is truly mind-blowing, and has given rise to a whole new branch of biomedical research, sometimes referred to as "systems biology." As described in Chapter 3, disturbances in this regulatory system are turning out to be more important than glitches in the proteins themselves as contributors to the risk of common disease.

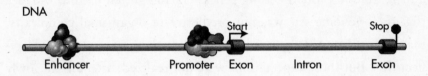

Figure B.1: A simplified cartoon of gene structure and function. The gene, made up of DNA, is transcribed into RNA beginning at a "start" signal. The DNA sequences just upstream of that, making up the "promoter," serve as recognition sites for RNA polymerase and a variety of other transcription factors, signaling the cell that this gene should be actively transcribed. Other DNA "enhancer" signals located some distance away often aid in this process. The initial RNA copy extends across the entire gene, but the introns are subsequently spliced out in the process of RNA maturation.

Associated with this study of the control of gene regulation has been the emergence of a new field, epigenetics. This term refers to the fact that the function of a DNA molecule derives not just from the sequence of bases, but also from the way DNA has been modified by other forces. For instance, the cytosine bases in DNA are often modified by the chemical addition of a methyl group during the lifetime of a cell, and that modification tends to shut down the function of the associated DNA. Similarly, binding of various proteins to the DNA molecule may render a particular stretch of DNA accessible or not accessible to the machinery that would generate a messenger RNA, and ultimately a protein, from that packet of information.

Much of this epigenetic marking of DNA is carefully controlled by developmental signals that are themselves encoded within the DNA instruction book. These marks can also be affected by environmental exposures. That explains why the study of epigenetics has become of such great interest to those trying to understand how heredity and environment interact to result in health or disease. And clearly there are interesting things to be discovered here.

Genetic Variation: Spelling Differences in DNA

As depicted in Figure 3.2, the DNA sequence is remarkably similar in any two individuals, but occasional differences occur. Most of these are single-letter variations, and are referred to as single nucleotide polymorphisms, or SNPs. The two alternative spellings of the SNP are referred to as alleles.

We humans are diploid—that is, we actually have two copies of the genome inside each cell. One copy is contributed by the sperm and one by the egg, coming together at the moment of conception. Thus, although we customarily say that the human genome is 3.1 billion base pairs, each cell actually has twice that amount of DNA. Being diploid also means that for a typical SNP that has an A and a T allele, an individual might have both copies as A (referred to as homozygous A), both as T (homozygous T), or one of each (a heterozygote).

There are about 10 million common SNPs in the human population (three are shown in Figure 3.2). Most of these were present in our common ancestors, a group of about 10,000 individuals who lived in eastern or southern Africa about 100,000 years ago. Because only 5,000 generations separate us from that founder pool, there has been insufficient time for genetic variation to become thoroughly scrambled in modern-day humans. The practical consequence is that SNPs tend to travel in chromosomal neighborhoods, and knowing what allele is present at one SNP on a chromosome often allows an accurate prediction of the alleles at SNPs nearby. Figure B.2 shows how this correlation between nearby variants has come about.

Ancestral Chromosomes

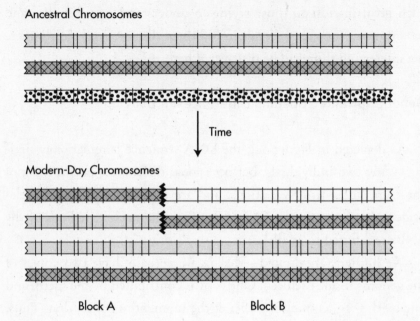

Figure B.2: The SNPs in the human genome tend to travel in neighborhoods, or "blocks." All humans are derived from a small group of about 10,000 ancestors. Three homologous segments of chromosomes that were present in these common ancestors are shown at the top; each vertical tick mark represents a SNP. Over the last 5,000 generations, the third ancestral type (stippled) was lost, but the other two were transmitted to modern times. In some copies (the bottom two), these segments appear identical to the ancestral versions, but in other copies a "hot spot" of recombination has allowed a crossover to generate two new chromosomal types. The SNPs within blocks A and B will remain tightly correlated with each other, but the correlation between SNPs in block A and block B will be limited by the presence of such crossovers. The HapMap project defined the boundaries of these blocks for selected European, Asian, and African populations.

The International HapMap Consortium was set up to define the boundaries of these SNP neighborhoods. Now that we know this, it is possible to study a small subset of the 10 million common SNPs and effectively survey the entire genome, since the SNPs chosen ("tag SNPs") serve as proxies for all the rest. That strategy has allowed the success of genome-wide association studies (GWAS; see Chapter 3) for many common diseases.

Large-Scale Variations in DNA

Another important revelation about the human genome relates to the long-range structure of DNA. Although regions of the genome that line up nicely show only about one or two differences in 1,000 base pairs between individuals, there is a fraction of the human genome that consists of long-range duplications of DNA sequence, extending over thousands of base pairs. These may vary in the number of copies between individuals. Many of these so-called copy number variations (CNVs; see Figure 3.7) occur in parts of the genome that seem to be depleted in known genes, but that is not true of all of them. Remembering that we each have two complete copies of the genome, one from each of our parents, you would normally expect that a particular sequence of DNA occurs exactly twice in an individual. But these copy number variations break that rule, and genes located within those segments can be present in only one copy, or not present at all, in some individuals, or in as many as six or eight copies in other individuals who have tandem duplications.

In some instances, this kind of copy number variation can lead to a medical condition. In the introduction, I mentioned my father-in-law, recently diagnosed with a neurodegenerative disorder called Charcot-Marie-Tooth disease. Dr. James Lupski, himself affected with this condition, demonstrated more than 10 years ago that it arises because of a duplication of about 1 million base pairs of DNA sequence containing the gene *PMP22*. Normally there are two copies of *PMP22*, but my father-in-law and others with Charcot-Marie-Tooth disease have three. The discovery of CNVs has led to some confusion about how to describe the similarities and differences between individuals at the genome level. For the part of the genome that is not involved in CNVs, the answer is that two unrelated individuals are 99.9 percent identical. But when we account for the CNVs, these two individuals are probably more like 99.6 percent identical.

For More Information

National Human Genome Research Institute at NIH: www
.genome.gov/Education

The DNA Learning Center at Cold Spring Harbor: http://www
.dnalc.org/

University of Utah Learn.Genetics Center: http://learn.genetics
.utah.edu/

Kansas University Medical Center Genetics Education Center:
http://www.kumc.edu/gec/

Gelehrter, T. D., F. S. Collins, and D. Ginsburg: *Principles of Medical Genetics.* Lippincott, Williams, and Wilkins, New York, 1998.

A Brief Personal History of the Human Genome Project

The Grail of Biology?

The rapid accumulation of knowledge about the human genome that is making personalized medicine possible can be traced directly to the Human Genome Project (HGP), considered by many to be one of the boldest scientific efforts that humankind has ever mounted. Conceived of in the late 1980s amid much controversy, the HGP got under way in 1990 and promised to read out all the letters of the human DNA code by 2005. Many saw that promise as wildly optimistic, since at the time DNA sequencing was slow, inefficient, and much too expensive to contemplate scaling up to a target of this size.

No less a figure than Jim Watson himself (half of the Watson-Crick team who discovered the double helix structure of DNA in 1953) was recruited to jump-start the HGP in the United States. He effectively sweet-talked critical members of Congress into providing start-up funds. Watson was a master at this, even intentionally rumpling his hair and untying his shoes before entering a congressional office, in order to cultivate the image of a goofy but lovable professor. Watson also persuaded a group of young scientists to take a chance on being part of this historic adventure. I was one of them, founding and leading a genome center in 1990 at the University of Michigan. But just as the U.S. effort was getting under way, a public dispute about gene patenting led Watson to resign. A cloud of gloom settled over the still young and struggling genome community. Who would step into Watson's legendary (but often untied) shoes?

No one was more surprised than I when the search settled on me. But the search committee and the director of the National Institutes of Health (NIH) thought that a genome scientist who was also a physician would be a good choice. My parents, having been burned by political manipulation when they worked for Eleanor Roosevelt in the 1930s, warned me that working for the government would be a big mistake, and I declined at first. But how could I walk away from the chance to lead such a historic enterprise? After a few months of internal struggle, I agreed.

When I arrived at NIH in 1993, I had a sinking feeling that I might have signed up to lead an enterprise that was doomed. The pathway to completion of the human genome sequence seemed hopelessly complex. Wisely, the planners of the genome project had laid out a series of milestones, including starting with more modest goals, but even those seemed extremely challenging. A critical goal was to improve the technology for DNA sequencing and test it on a few simpler organisms like bacteria, yeast, roundworms, and fruit flies, carefully chosen to tell us something about the basics of molecular biology. Some of the best and brightest biologists, geneticists, chemists, and physicists in the United States, the United Kingdom, France, Germany, and Japan (China joined later on) began to apply their talents to solving these problems, but it was not until 1996 that we gained enough experience to begin to contemplate extending beyond these simpler model organisms. In that year, we gathered at an international meeting in Bermuda to talk about actually piloting the sequencing of human DNA. This meeting was important, as it solidified the need for international cooperation. But no one was willing to argue at that point that the 2005 deadline was realistic.

Perhaps even more important, this was the meeting at which all present agreed that the DNA sequence of the human genome was of such fundamental importance that it should be immediately released onto the Internet every 24 hours, not subjected to secrecy or patenting, and not even kept waiting for publication in a scientific journal. The assembled scientists, many aware that they might not actually have the authority to speak for their host country, unanimously endorsed a resolution that "all human primary genomic sequence from large-scale sequencing centers should be freely available and in the public domain for both research and development, in order to maximize its benefit to society."

Concerned that a gold rush toward patenting human DNA, already under way by a few companies and universities in 1996, might blunt the utility of this information for public benefit, the assembled group also endorsed this statement: "The genomic sequence, in the absence of any experimental information about functional or diagnostic use, is not an appropriate subject for patent protection." These bold statements about open data access were radical at the time. Looking back, one might legitimately say that there were two major contributions of the human genome project to human health: the first was the sequence itself, and the second was the decision about free and open access. That principle has now spread to many other areas of biomedical research, accelerating progress and the ultimate benefit to the public.

Although the HGP was from the start an international effort, the United States had the largest single investment. As the leader of the

U.S. effort, I found myself in the role of overall project manager. With 20 centers in six countries involved in the work, this was a rather massive challenge, especially with the need to maintain consistent production schedules and high-quality data. Most biomedical researchers had not previously been involved in a "big science" activity of this sort, and many adjustments had to be made by the senior project leaders, many of whom had substantial egos and were used to running independent research laboratories.

But the chance to read the human DNA sequence for the first time was a scientific challenge in the same category as splitting the atom or going to the moon. One could even argue that the HGP was even more significant than those other achievements, as this adventure was devoted to exploring ourselves, with an opportunity for unprecedented health benefits for humankind. Motivated by that shared vision, and linked by electronic communication, huge numbers of conference calls, and regular exchange of personnel, the 2,500 scientists in 20 centers coalesced into a coordinated and integrated team.

Public or Private?

Then a large cloud appeared on the horizon. The maverick scientist Craig Venter, supported by the deep pockets of the Applera Corporation, announced in May 1998 that he had access to $400 million and would mount a massive competitive private effort to read out all of the human genome sequence.

Though the business plan was initially unclear, it soon emerged that Dr. Venter and his company, Celera, planned to recoup the stockholders' investments by patenting an unspecified number of genes, and by charging subscription fees for those who would like to look at the DNA sequence data. Celera also planned a somewhat different technical approach to determining the DNA sequence. Whereas

the public genome project was pursuing a strategy that involved assembling several pages of the book of life at a time, Celera aimed to
obtain random sentences of the code in very large numbers, and then
to assemble the whole thing computationally. That latter strategy had
worked brilliantly in Venter's hands for simpler genomes, but was of
uncertain promise for a genome of the complexity of the human, especially given all those repetitive sequences that could potentially choke
the assembly process.

The next two years were tumultuous. Some observers were enamored of Celera's approach, and even argued that the human genome
project should be completely privatized. But the public project sol-
diered on, ramping up energy and capacity to the point where 1,000
letters of the DNA code were being generated every second, seven days
a week and 24 hours a day, releasing the data immediately each day.
Celera also demonstrated its technical muscle, but its business model
meant that none of its data were available for scrutiny. Celera soon
realized that the daily release of information from the public project
could be usefully merged with its own information. Ultimately, as a
consequence of this free DNA data, Venter stopped far short of his
own production goals, and more than half of the "Celera assembly"
ultimately turned out to be downloaded data from the public project.

Nonetheless, by May 2000, as I was addressing the assembled genome science community at Cold Spring Harbor (see Chapter 1), it
was time for a truce. The so-called race for the human genome was
becoming unseemly, and threatened to detract from the real goal of
the project: improving human health. Working with my friend Ari
Patrinos of the Department of Energy, I had a few secret meetings
with Venter, and a joint announcement was arranged.

On June 26, 2000, Venter and I stood next to Bill Clinton, the
President of the United States. Leaders of the scientific and medical
communities joined us in the East Room of the White House as the

world's press watched, and we were even linked by satellite to Prime Minister Tony Blair in the United Kingdom. Comparing the HGP to Lewis and Clark, Clinton spoke these words: "Today the world is joining us here in the East Room to behold a map of even greater significance. We are here to celebrate the completion of the first survey of the entire human genome. Without a doubt, this is the most important, most wondrous map ever produced by humankind. Today we are learning the language in which God created life. We are gaining ever more awe for the complexity, the beauty, the wonder of God's most divine and sacred gift. With this profound new knowledge, humankind is on the verge of gaining immense new power to heal. Genome science will have a real impact on all our lives—and even more, on the lives of our children. It will revolutionize the diagnosis, prevention, and treatment of most, if not all, human diseases."

Well, you might say, politicians are prone to exaggeration, so perhaps these stirring sentences might be considered a bit overblown. But listen to the words of Matt Ridley, written only a month after the White House event, and appearing in the opening lines of his compelling book *Genome*: "I genuinely believe that we are living through the greatest intellectual moment in history. Bar none. Some may protest that the human being is much more than his genes. I do not deny it. There is much, much more to each of us than a genetic code. But until now human genes were an almost complete mystery. We will be the first generation to penetrate that mystery. We stand on the brink of great new answers, but, even more, of great new questions."

Shortly after this, realizing that complete public availability of the DNA sequence made its business plan no longer viable, Celera changed directions, let Dr. Venter go, and went on to become a diagnostic company. Meanwhile, the International Human Genome Sequencing Consortium continued its work. Proving the principle that the last 10 percent of a production project takes almost as much effort as the first 90 percent, the final completed sequence of the human genome was announced in April 2003. The celebration came almost exactly 50 years after Watson and Crick's description of the double helix.

Today, virtually all observers agree that the complete and immediate public availability of the human genome sequence was a critical component of its success. And yet the outcome could easily have been different. Had the cries for privatization of this effort won out in 1999, this would now be a very different world.

Further Reading

Ridley, M. *Genome. The Autobiography of a Species in 23 Chapters.* New York: HarperCollins, 1999.

Shreeve, J. *The Genome War. How Craig Venter Tried to Capture the Code of Life and Save the World.* New York: Knopf, 2004.

Sulston, J., and G. Ferry. *The Common Thread: A Story of Science, Politics, Ethics, and the Human Genome.* Washington, DC: Joseph Henry, 2002.

APPENDIX D

Rational Drug Development

In Chapter 2, the pathway to development of new drugs was touched on, but the actual steps involved in going from an understanding of the molecular basis of disease to an FDA-approved treatment were not spelled out. This is a complex area of science, but some readers will be interested in learning more. Appendix D will review briefly the basics of how new drugs for rare and common diseases are being developed.

Drugs Are Chemicals

First of all, we need to consider what kinds of molecules drugs are. Most drugs are organic compounds made up of carbon, nitrogen, oxygen, hydrogen, and occasionally other atoms, structured into a particular shape that allows them to interact with a human protein to either enhance its function (we often call this category an agonist) or inhibit its function (usually called an antagonist). Many older drugs, such as aspirin (Figure D.1), were arrived at purely empirically, without any idea of their mechanism of action until much later. More recently, many drugs have been developed by identifying a particular protein as a target, and then screening large numbers of extracts from natural sources, such as fungi, to try to find an activity of a desired sort. A prime example is the development of the first statin drug, identified by its ability to block a critical step in cholesterol synthesis (also shown in Figure D.1). Statins are now the most widely prescribed drug in the United States, and have prolonged many lives because of their ability to prevent coronary artery disease and heart attack.

Aspirin   Lovastatin

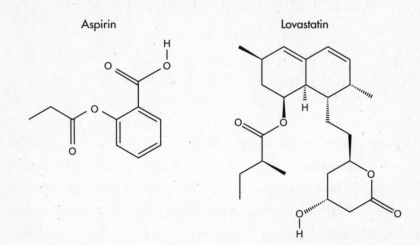

Figure D.1: The molecular structures of two common drugs: aspirin and lovastatin. These standard chemical drawings involve some shorthand conventions that are familiar to chemists—a carbon atom is inferred to be located at all the apexes of each structure, hydrogen atoms are left out in most instances, single lines indicate a chemical bond, and double lines indicate a double bond.

Systematic Searching for New Drugs

More recently, the drug development process has shifted toward a more comprehensive strategy: identifying organic compounds that might have the desired activity from a synthetic collection of pure substances, rather than depending on natural sources. A "library" of hundreds of thousands of these compounds, which can be thought of as a collection of "shapes" in "chemical space," can then be tested against a particular target. Evidence of activity of a particular compound, or "small molecule," as it is often called (to distinguish it from "large molecules" like proteins or monoclonal antibodies), can be followed by a successive series of chemical modifications until an optimal structure is identified.

Going back to the specific example highlighted in Chapter 2, this very strategy has been vigorously pursued for cystic fibrosis (CF), with unprecedented scientific and financial support from the Cystic Fibrosis Foundation. First of all, it was necessary to come up with a simple scheme, known as an assay, that could identify a few compounds out of hundreds of thousands that might show potential for correcting mutations in the *CFTR* gene. As it was known that CFTR functions as a chloride ion channel, pumping chloride from the inside of the cell to the outside, a clever method was implemented, using a fluorescent dye that was sensitive to the concentration of chloride inside the cell. In uncorrected cystic fibrosis cells, growing in a laboratory culture dish, the chloride concentration will remain high and the fluorescence will be bright, even after a stimulus that would normally open the CFTR channel (Figure D.2). Hundreds of thousands of compounds were tested in this assay, and a few were identified that appeared capable of reducing the fluorescence. Those four became the starting point for a highly creative program of drug development.

But now, the consequence of genetic heterogeneity—or of multiple disease alleles, to use the genetic terminology—becomes important for understanding treatment options. You see, not all of those 1,000 different *CFTR* mutations have the same molecular mechanism. The common ΔF508 mutation results in a protein that fails to fold properly, gets hung up in traffic, and never reaches the cell membrane to carry out its chloride transport job. Certain other *CFTR* mutations, exemplified by one called G551D (a mutation that changes glycine to glutamic acid at amino acid 551), do not have the folding and traffic problem, but the protein still doesn't function properly in its location in the membrane.

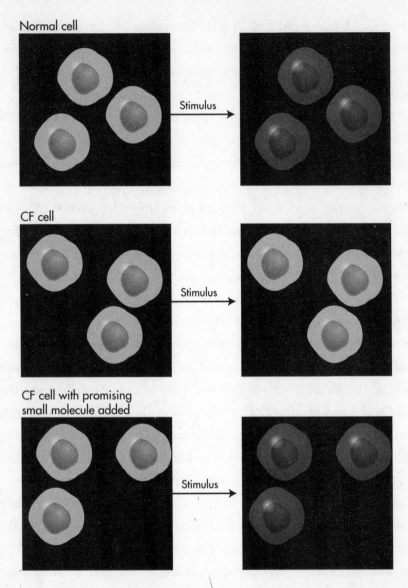

Figure D.2: A fluorescent assay for compounds that might be beneficial in cystic fibrosis (CF). To begin with, cells are loaded up with a fluorescent dye that senses the presence of chloride ions. Then a stimulus is delivered that normally opens the CFTR chloride channel, causing chloride to flow out, and reducing the fluorescent signal inside the cell (top panel). Lacking functional CFTR, a CF cell will remain brightly fluorescent even after the stimulus (middle panel). If a small molecule capable of activating the channel is present, however, the fluorescent signal will fade from the cell after the stimulus (bottom panel). Millions of compounds were tested to find a few that had this property.

It should be apparent that these different molecular problems will require potentially different drug strategies. Accordingly, the drug development plan for CF has two arms: in one component, a drug called the "corrector" is being developed that helps with the protein folding and traffic problem. This will be needed to treat ΔF508. Second, a different compound, called the "potentiator," is being developed that helps the protein function once it gets to its proper location in the cell membrane. This should work for G551D. That is the drug that Bill Elder received.

The Long Path to Approval for Human Use

Finding a small molecule that appears to correct the disease defect in a test tube is an exciting development, but many steps still lie ahead. To be useful in humans, the compound must be absorbed (preferably by mouth, so that injections are not needed); it must reach levels high enough in the relevant tissues to have a therapeutic effect; it must have a reasonable half-life in the body (so that doses don't have to be given more than four times a day at most); and it must not be toxic. The preclinical phase of drug development aims to optimize all these features at once, using animals for testing to avoid exposing humans to potential toxic outcomes. These steps often take several years, and 95 percent or more of compounds fail at this point.

Once a compound appears to have all the right properties in an animal testing system, application is made to the FDA for permission to try it in humans. If that is granted, the first test is usually done with a low dose on a small number of normal volunteers who have consented to this research, to look for any evidence of unexpected toxicity. This is called phase I. If all goes well, a phase II trial on dozens to hundreds of individuals with the disease is initiated, to look for evidence of benefit and to define the optimum dose. If that study also proves successful, a multicenter phase III trial on hundreds to thousands of patients is conducted. The gold standard in such trials includes randomizing the participants to receive either the new therapy or the previous standard, with neither the patients nor the doctors knowing which treatment they are receiving. Without this double-blind feature, therapeutic trials can be misleading, as hopeful researchers and patients may sometimes conclude that a treatment has been beneficial when chance alone could account for it. Careful analysis of outcomes is done to look for any evidence of unexpected complications. In the modern era, all patients should also have extensive DNA analysis conducted, so that subsets with unusually good or bad responses can be identified.

If the phase III trial shows clear benefit and acceptable risk, the drug manufacturer can then apply to the FDA for approval to market the drug as part of general medical care. Generally, however, the FDA will require at least two independent phase III trials before granting approval.

Rare and Neglected Diseases

The process described above takes many years, costs hundreds of millions of dollars, and has a high failure rate. Biotechnology and pharmaceutical companies quite understandably are reluctant to make that kind of investment for diseases that have limited market potential. That includes the more than 6,000 diseases that are considered rare, and also diseases of the developing world that are actually quite common. As a result of this economic reality, development of new treatments for these diseases has been slow or nonexistent. Recently, however, initiatives have been mounted in the government and philanthropic sectors to change that, by providing tools and technologies to enable academic investigators to play a more significant role in drug development, thereby "de-risking" projects for eventual adoption by the private sector. The progress in drug development for cystic fibrosis, largely paid for by private donations to the Cystic Fibrosis Foundation, is an excellent example of how this approach can work for an uncommon disease.

Services Provided by
Direct-to-Consumer Genetics Companies
Offering Broad-Spectrum Testing

(as of May 2009)

DISEASES

Condition	23andMe[1]	deCODE	Navigenics
Abdominal aortic aneurysm	(+)	+	+
Alzheimer's disease		+	+
Asthma	(+)	+	
Atrial fibrillation	(+)	+	+
Basal cell carcinoma	(+)	+	
Bladder cancer	(+)	+	
Celiac disease	+	+	+
Chronic lymphocytic leukemia	(+)	+	
Colorectal cancer	(+)	+	+
Crohn's disease	+	+	+
Essential tremor	(+)	+	
Exfoliation glaucoma	(+)	+	+
Gallstones	(+)	+	
Gout	(+)	+	
Graves' disease			+
Heart attack	(+)	+	+

	23andMe	deCODE	Navigenics
Intracranial aneurysm	(+)	+	+
Lung cancer	(+)	+	+
Lupus	(+)		+
Macular degeneration	+	+	+
Melanoma	(+)		+
Multiple sclerosis	(+)		+
Obesity	(+)	+	+
Osteoarthritis			+
Parkinson's disease	+		
Peripheral arterial disease	(+)	+	
Prostate cancer	+	+	+
Psoriasis	+	+	+
Restless legs syndrome	(+)	+	+
Rheumatoid arthritis	+	+	+
Sarcoidosis			+
Stomach cancer	(+)		+
Thyroid cancer		+	
Type 1 diabetes	+	+	
Type 2 diabetes	+	+	+
Ulcerative colitis	(+)	+	
Venous thromboembolism	+	+	+

TRAITS

Trait	23andMe[1]	deCODE	Navigenics
Alcohol flush reaction	+	+	
Ancestry	+	+	
Bitter taste perception	+	+	
Earwax type	+		
Eye color	+		
HIV/AIDS resistance (CCR5)	+		
Lactose intolerance	+	+	+
Malaria resistance (Duffy)	+		

Male pattern baldness	(+)	+
Muscle performance	+	
Nicotine dependence	(+)	+
Non-ABO blood groups	+	
Norovirus resistance	+	

DRUG SENSITIVITY

Drug	23andMe[2]	deCODE	Navigenics
Clopidogrel (Plavix)	+		
Coumadin (Warfarin)	+	+	

CARRIER STATE

Disease	Carrier at Risk?	23andMe	deCODE	Navigenics
Alpha–1–antitrypsin deficiency	Only smokers	+		
BRCA1/BRCA2	Yes[3]	Selected[4]		
Bloom's syndrome	No	+		
Cystic fibrosis	No	ΔF508[5]		
G6PD deficiency	Yes[6]	+		
Glycogen storage disease Ia	No	+		
Hemochromatosis	No	+	+[7]	
Sickle-cell Anemia	No	+		

---

1. For 23andMe, conditions and traits marked (+) are classified as "Research Reports": that is, the company believes no consensus about the significance of a genetic test result has been reached yet. DeCODE does not make this distinction. In addition to the entries shown in this table, 23andMe reports on 35 conditions and 18 traits in the "Research Reports" category, for which neither deCODE nor Navigenics reports results.
2. 23andMe includes three additional predictions of drug sensitivity in the "Research Reports" category, for which deCODE does not report results.

3. Women with *BRCA1/2* mutations are at high risk for breast and ovarian cancer; men are at slightly increased risk for prostate, pancreatic, and male breast cancer (see Chapter 3).

4. Test only detects three *BRCA1/2* mutations that are more common in Ashkenazi Jews. A negative test therefore does not rule out the possibility of other mutations in *BRCA1/2*.

5. Test only detects the ΔF508 mutation in *CFTR* (see Chapter 2), so a negative test does not rule out the possibility of being a CF carrier.

6. Generally, only male G6PD carriers are at risk for hemolytic anemia after fava beans or certain drugs, since the gene is on the X chromosome.

7. deCODE detects hemochromatosis carriers, but doesn't report them as such.

INDEX

Entries in *italics* refer to tables and illustrations.

DRD2/DRD4 gene, 200
drug addiction, 185–86
drug-metabolizing enzymes, 234–35, *241*
drugs. *See also specific diseases and drugs*
  adverse reactions to, 231–34, 239–44, 247–48, 250
  designer, 36, 134–35
  development of, 36–37, 307–13, *308*
  diabetes and, 72, 78
  dosage and effectiveness of, 181, 231–47, *235*, *241*, 250
  FDA approval of, 137, 311–12
  heart disease and, 79
  HIV/AIDs and, 168–69
  longevity and, 217
  macular degeneration and, 68
  malaria and, 173
  Marfan syndrome and, 40–41
  names of, 235–36, *236*
  over-the-counter, 20
  personalized medicine and, 272, 274
  pharmacogenomic revolution in, 247–50
  progeria and, 227
  race and, 160–62
Druker, Dr. Brian, 129–32
Duchenne muscular dystrophy gene, 7
Dulbecco, Renato, 105
Dutch ancestry, 152
Dx-Rx paradigm, 239, 242, 245, 248

eczema, 179–80
EGFR gene, 136–37
EGFR growth-factor receptor, 246
Eisenhower, Dwight D., 239
Elder, Bill, 36, 311
electrocardiograms (EKGs), 18
electronic medical records, 15, 233, 275–76
embryo selection, 54–57
emphysema, xxiii, 123, 140
employers, 92, 115–17
Encyclopedia of DNA Elements (ENCODE), 12
endocrine system, 118
endometrial samplings, 120
endoxifen, 247

Energy Department, 132
environmental factors, xxiv–xxv, 11, 14, 20, 28, 32, 37–41, 56, 61–62, 67, 78, 93–94, 109, 123–26, 156–58, 169–70, 185–88, 195–96, 202, 215, 221, 223, 294–95
enzymes
  defined, 283
  drug metabolism and, 234–35, *235*, 237, 240–41
epigenetics, 294–95
Epstein-Barr virus, 126
ERBB2 gene, 128
Erbitux. *See* cetuximab
esophageal cancer, 104, 123
estriol screening, 51
estrogen antagonists, 246
estrogen receptors, 246–47
ethnic ancestry, xvi, xxiii, 10–11, 84, 91, 142–53, *148*, 164. *See also* race *and specific ancestry*
eugenics movement, 201
European ancestry, 91, 145, 150–51, 156–58, 171–72
evolution, 166–67, 215–16. *See also* natural selection
exercise, xx, 62, 77–78, 89, 91, 96–97, 226, 269, 271–72, 274
exon, 9, 283, 286, *294*
ex vivo gene therapy, 253–55, *254*, 267
eye color, xxiii, 55
eye tumors, 118

familial adenomatous polyposis (FAP), 118–19
family 15, 99–104, *100*, 108–10, 113, 115
family health history, xviii–xix, xxii, xxv, 14–21, 58, 61, 90, 119–20, 141, 226, 268–69
Family Health History Initiative, 14–15, 22
Fanconi anemia, 56–57
farnesyl transferase inhibitor (FTI), 227–28
Federal Bureau of Investigation (FBI), 153
Federal Trade Commission (FTC), 94
fibrillin gene mutation, 41